BRIAN CLEGG was born in Lancashire. He graduated in Natural Sciences from Cambridge University and gained a second MA at Lancaster University. Brian worked with Dr Edward de Bono before setting up a consultancy whose clients include the BBC, Sony, HM Treasury and the Met Office. His books have been translated into several languages, and include *A Brief History of Infinity: The Quest to Think the Unthinkable* and *Build Your Own Time Machine: The Real Science of Time Travel* (published by Duckworth Overlook). He has lectured at the Royal Institution in London and is a Fellow of the Royal Society of Arts. Brian lives in Wiltshire with his wife and twin children.

ALSO BY BRIAN CLEGG

How to Build a Time Machine

Inflight Science

Armageddon Science

Before the Big Bang

Ecologic

The God Effect

A Brief History of Infinity

The Man Who Stopped Time

The First Scientist

Light Years

GRAVITY

HOW THE WEAKEST FORCE IN THE UNIVERSE SHAPED OUR LIVES

BRIAN CLEGG

Duckworth Overlook

This paperback edition 2013
First published in the UK in 2012 by
Duckworth Overlook
30 Calvin Street, London E1 6NW
T: 020 7490 7300
E: info@duckworth-publishers.co.uk
www.ducknet.co.uk
For bulk and special sales, please contact
sales@duckworth-publishers.co.uk
or write to us at the address above

First published in the USA in 2012 by
St. Martin's Press, New York

A catalogue record for this book is available from the British Library

ISBNs
Paperback: 978-0-7156-4407-2
Kindle: 978-0-7156-4446-1
ePub: 978-0-7156-4445-4
Library PDF: 978-0-7156-4444-7

Manufactured in Great Britain by
CPI Group (UK) Ltd, Croydon, CR0 4YY

FOR GILLIAN, REBECCA, AND CHELSEA

CONTENTS

ACKNOWLEDGMENTS

My grateful thanks to my editor, Michael Homler, for his guidance and support.

Thank you to the many people who have helped me along the way to writing this book, including the staff of the Science Museum Library (Wroughton), Dr. Marcus Chown, Professor Brian Cox, Professor Bob Evans, Dr. Sean Carroll, Professor Peter Haynes, Professor Friedrich Hehl, Professor Michio Kaku, Professor Ronald Mallett, Dr. Peet Morris, and Professor Günter Nimtz.

GRAVITY

CHAPTER ONE

WHAT GOES UP

> *So in all their procedinges . . . they shew themselffes to be men of gravyte and wisedom.*
>
> —*State Papers of Henry VIII* (1849), VII.614

Hold this book in your hand and let go. What will happen? It's such an obvious question that it feels embarrassing to have to ask it. But humor me. What will happen? You don't have to carry out the experiment to know the answer. The book will fall. Why? Just as embarrassingly obvious. Because of gravity.

This is the most directly obvious force of nature. Its influence is programmed into our expectations of the world around us. If we let go of something and it drifts upward instead of falling, it's a double-take moment. Either we're dealing with something special, like a helium balloon, or we're not firmly planted on the Earth. When we drop things, they fall, simple as that. And yet, as we will discover, the story of gravity is anything but simple.

Gravity is so familiar and apparently obvious that we often miss seeing just how remarkable it is. Most rational people laugh at the idea of astrology. They may tolerate it as fun, but they accept that it is garbage. It's bizarre, they say, that anyone should believe that our lives are influenced in any way by astronomical bodies that are millions of miles away. Yet we accept that gravity—an invisible force with no detectable mechanism for exerting an influence—can have a real effect across just such distances. After all, the only thing that keeps the Earth in orbit is the gravitational attraction between it and the Sun, 93 million miles away.

You will sometimes see this distant reach of gravity being used to try to give astrology a scientific basis. We are subject to the gravitational attraction of the planets, the argument goes, so they *can* have an influence on our lives. While this is strictly true, it is worth bearing in mind that the gravitational force between a human body and the distant planets is tiny. By comparison, the gravitational attraction between a baby and the midwife is greater. So if astrology really were based on this idea, we should have astrological charts including the position and mass of the midwife, and everything else that was present at the birth.

In the real world of science, gravity has a much greater effect on us than anything astrologers could even imagine. Without gravity there are just so many ways that we wouldn't exist or be able to carry out our everyday activities. It isn't just a convenient way of sticking to the surface of the Earth.

It is thanks to gravity that bodies like planets and stars came into existence in the first place. Just imagine you are visiting the site of the solar system before it formed, around 4.5 billion years ago. You are looking at a cloud of matter—gas and dust floating in space. There is no wind to disturb this collection of material, so it will not be blown from place to place, but there is gravity. Each of the specks of matter has a tiny influence on the others. Gradually, painfully slowly, the matter will be pulled together.

At the same time the whole thing is rotating. That's the way it started out, and there is nothing to stop it. So as the matter bunches together, it is also whirling around, like the disk of a pizza as the dough is spun between the hands of the baker. Eventually, at the center of this whirling cloud will be a large clump of matter. As each new particle comes crashing in, it will add energy, producing heat. (Think of the way rubbing your hands together produces heat. It's a much smaller effect for each particle but there are many billions of particles contributing.)

After millions of years of collecting particles in this central lump there will be enough heat and pressure from the gravitational pull of the accumulated mass for something remarkable to happen. Hydrogen atoms (or to be more precise hydrogen atoms each stripped of its electron to leave hydrogen ions, simple protons) will be forced closer and closer together with more and more energy. Eventually a reaction will occur. In a multistage process the hydrogen nuclei fuse to form helium, the next element up the chemical chain. In this process energy is released.

The released energy from the nuclear fusion gives even more oomph to the reaction, sending it snowballing through the mass of the central lump. What we are seeing is a star being kickstarted. The fusion process is the power source of the star. Without gravity, this could never have happened.

As described above, the process has one flaw. The positively charged hydrogen ions do not like to get near to each other. The closer you push them, the more the electrical charge fights back. The electromagnetic force causing this repulsion is much stronger than gravity. Even all the gravitational pressure of a star, plus the heat that has built up, is not enough to force the positively charged ions close enough together to fuse.

The final hurdle is overcome by quantum effects. Just as general relativity—the theory that explains gravity—deals with the very large, quantum theory explains the behavior of the very small. One of the oddities of quantum particles, like hydrogen nuclei, is that they don't have a specific position. They just have a range of probabilities as to where they might be. So although a pair of hydrogen nuclei are most likely to be held too far away to fuse by repulsion, this quantum uncertainty enables some particles to perform a process called quantum tunneling.

The particles have a small probability of finding themselves on the other side of the gap separating them without traveling through the space in between. Although the chances of any particular particle undergoing tunneling are very low, there are so many particles in a star that vast quantities of them make this jump every second. The Sun, for example, converts around 4 million tons of matter into energy every second, all derived from the minute difference in mass that arises when particles fuse.

In a normal star there is a balance, an equilibrium between the inward gravitational pull that drags all the particles in the star toward its center, and outward pressure. This pressure is a combination of traditional gas pressure—the result of the gas particles in the star bouncing off each other and resisting collapse—and the pressure of the light emitted in the fusion process. This reaction is going on deep in the Sun. When mass converts to energy it comes out as light. But the light doesn't come straight out—far from it.

Any particular photon of light will only travel a tiny distance before colliding with another matter particle and being absorbed. The light is then reemitted. This process acts as if the light were another particle that has bounced off the matter. As a result it gives some energy to the matter. The reemitted photon is slightly lower energy and the extra energy of the matter particle results in extra pressure resisting the collapse of the star.

As the photons very gradually make their way out of the star, they reduce in energy all the way. This is reflected in the differences in temperature through the cross section of a star. The Sun, for example, has a core temperature of around 10 million degrees Celsius (18 million °F), while the outer layer that we see is only around 5,500 degrees Celsius (9,900 °F). There are so many absorptions and reemissions along the way that photons take somewhere between 10,000 and 1 million years to get out of the Sun.

Meanwhile, back at the newly formed solar system, other clumps, whirling around that young star will also be coalescing under the attractive force of gravity, producing the planets. Not all will succeed though. This is a complex interaction. If there are enough big planets nearby, the small bits of matter might never have a chance to coalesce into a planet. This is thought to be the source of the asteroid belt between Mars and Jupiter.

Once assumed to be the wreck of a planet it is now thought to be pre-planetary material that never made it because of the disruptive gravitational fields of its neighbors.

All thanks to gravity. This omnipresent force might be a pain when we drop something or fall, but without it there would be no Earth. Even if gravity was somehow switched off after the formation of the planets, we still couldn't live. Apart from anything else, the only thing that keeps the atmosphere in place is gravity—not to mention keeping our feet firmly planted on the Earth's surface. You only have to watch the difficulties that astronauts on the International Space Station have undertaking basic tasks (including the familiar bodily functions) to realize that gravity is beneficial for everyday existence.

More subtly, if people stay without that familiar gravitational pull for too long, their muscles begin to waste and their bones deteriorate. All the evidence is that we couldn't exist for a full lifetime in a weightless state. Evolution has developed us (and all the plants and animals around us) to work under the influence of gravity—it is as essential for our long-term existence as the air that we breathe.

If plants are grown in space, the roots head off randomly, struggling to find nutrition. Roots make use of the directive force of gravity to know which way to head, something that Charles Darwin was aware of, and that is easily demonstrated by turning a plant pot on its side and seeing how the roots grow. In zero gravity, plants get confused. It's even worse for birds' eggs. In an experiment on the space shuttle Discovery, bizarrely sponsored by the fast-food company KFC, it was discovered that a series of quail eggs failed to hatch without gravity to keep the yolks near the shell.[1]

No animals have lived their entire lifespans in space, but there are reasons to be concerned if humans were ever brought up in weightless conditions. Without gravity pulling down on the internal organs, lung capacity is reduced by a change in position of the diaphragm, while the liver floats higher in the body cavity leaving even less room for the lungs to function. A baby born in space may have seriously compromised ability to breathe, which combined with bone deterioration emphasizes just how much gravity is part of our natural environment.

It is impossible to escape the physical impact of gravity without traveling to space or enduring a zero-gravity flight, but even if we ignore its physical omnipresence, gravity still makes itself known in the way that it threads through our consciousness. The very word "gravity" stretches beyond a mere attractive force of nature. In my dictionary, the first definition of gravity is about being grave, serious, weighty, and important.

Looking back over the use of the word, this figurative sense provided its first noted use in English. It wasn't until the seventeenth century that the scientific meaning of gravity as that mystical attractive force became common. The earliest scientific use of "gravity" contrasts two concepts of ancient Greek physics, gravity and levity. Gravity was a tendency to move downward toward the center of the Earth, and hence of the universe. Levity was a tendency to move up and away from the center. Solid matter had gravity, airy things had levity. (Of course, such gravity was referred to earlier than the seventeenth century, but it would usually have been in Latin.)

Strangely, while gravity was first used figuratively in English, levity seems to have started with the physical meaning and then moved into its more commonplace use, meaning something that is light, frivolous, and quite possibly funny. The descriptive opposite of the solid, meaningful, and serious gravity.

Gravity seems to have exercised the minds of writers before artists. The earliest art on the walls of prehistoric caves appears to ignore the effects of gravity, allowing figures to float in space at will. But long before the constraints of perspective were imposed to give paintings and other illustrations a pseudo three-dimensional appearance that better reflected how we see the world, painted feet had become planted firmly on solid ground—gravity was there by implication.

So it would remain. In the work of surrealist painters like Salvador Dali, gravity sometimes played a more visible role, acting to distort the shape of familiar objects, but largely it would be an unconscious inclusion in the arts. Only when cinema needed to portray adventures in space and made its own attempts, all too often inaccurate, to portray the impact of differing gravitational force away from the surface of the Earth, would it be considered more explicitly.

To understand the origins of the idea of gravity in the scientific sense, though, we need to return to the ancient Greeks with their concepts of gravity and levity. To minds like ours, brought up on a Newtonian picture of existence, the Greek view provides us with a surprisingly alien perspective.

CHAPTER TWO

A NATURAL TENDENCY

> *Were it not for gravity one man might hurl another by a puff of his breath into the depths of space, beyond recall for all eternity.*
>
> —*Philosophiae Naturalis Theoria* (1758)
>
> Roger Joseph Boscovich

Picture a very simple world. A flat, uniform plane, featureless, that stretches from horizon to horizon. We put a person on this plane. They can walk along quite happily. Why not? Why should they float away from it? Anyone with a basic scientific training might be thinking, "Yes, but according to Newton's third law every action has an equal and opposite reaction. So without gravity, when you press your foot down on the ground, the ground would push you away and you would shoot off in the opposite direction."

This is true. But I want to take you back to a world before Newton, before our present-day assumptions about forces and reactions, to build the picture of the Greek idea of gravity. Difficult though it is, we need to put aside any assumptions based on Newtonian physics, to take a wide-eyed, open view.

So in this picture, our person on the flat-plane world could walk along without any problem. But now she takes an apple out of her pocket and lets it go. What happens to it? In this simplest of worlds it should just stay there. And it is hard to imagine why it should head straight down. If the apple did fall, we would just have to say, "That's the way things are," and not attempt an explanation. It would come down to, "That's what things do."

To be fair, the Greeks, unlike some other early cultures, did not have such a simplistic model of the universe. They knew that the Earth was ball-shaped rather than flat. There were two big clues that highlighted this. One was the way that the stars changed position in the sky depending on your location. The other was the natural viewpoint of seafarers. It doesn't take much experience at sea to realize that other ships and landmasses rise up over the horizon—whichever direction they come from. If you go up to the top of the mast, they appear sooner. It all suggests a curvature that seems to be the same in all directions, strongly hinting at a ball-shaped world.

Exactly when the Greeks firmed up the theory of the spherical Earth is not clear. It has been attributed to Pythagoras, but there seems little evidence for this.[1] The majority of the achievements we tend to associate with Pythagoras (including his famous theorem) don't seem to have been his original idea.[2] Instead they have picked up his name in a Greek tradition that required all great concepts to be the work of a Greek sage. (Pythagoras's theorem, for example, was in use a good 1,000 years before that philosopher's birth.)

What is clear is that by around the fifth century BC there was little argument about the shape of the Earth. Plato, working around 400 BC, could readily assume that our planet was spherical. "The Earth," said Plato, "when looked at from above, is like one of those balls which have leather coverings in twelve pieces, and is of divers colors, of which the colors which painters use on Earth are only a sample. But there the whole Earth is made up of them, and they are brighter far and clearer than ours. . . ."[3]

The third-century BC philosopher Eratosthenes, who is considered the first to treat geography as a discipline in its own right, managed an early measurement of the circumference of the Earth (a meaningless concept had it been flat). This was worked out by combining the distance between two points on the Earth (estimated by the time taken to travel between the two by camel), and the relative angle of the Sun at midday in the two locations. There was some luck in the relative accuracy of his measurement, yet the principle was good and there was no doubt that he came up with a fair order-of-magnitude size for the Earth.

At the time, the general understanding was that the Earth was fixed at the very center of the universe. Around it traveled the Moon, the Sun, and the planets, and finally the stars, fixed to crystal spheres that transported them through the heavens. And it is within this picture of the universe, with the Earth firmly in place at the heart of those rotating spheres, that the Greek concept of gravity evolved.

The principle that lay behind the Greek understanding of gravity was the nature of the elements. It was thought that just four elements made up everything in the world—earth, air, fire, and water—and that each had a natural tendency of behavior. Earth wanted to move toward the center of the universe. Fire wanted to get away from it. Air and water sat between the two, with air tending toward fire and water toward earth.

These four elements had been devised by Empedocles (born around 492 BC) and were set in concrete at the heart of philosophy for 2,000 years by Aristotle.[4] Again, we need to avoid modern modes of thought and accept for the moment a picture of the world built on those four elements.

As the planet Earth was thought to be at the center of the universe, that's where the heavier elements headed, while the lighter ones tried to get away, moving toward the skies. All of this behavior only applied beneath the crystal sphere carrying the Moon. Further out, a different element, the "quintessence," was in charge. This wasn't of Earthly nature and didn't have either tendency, so stayed forever in its place, fixed and unchanging. So strong was the belief that everything above the Moon was unchanging that it was felt that comets had to be beneath the Moon, or sublunary, because they came and went unpredictably.

This hugely flawed idea, powerful in its simplicity, largely held through into medieval times—though an increasing awareness of the difficulties of this model of the universe began to throw such a mechanism for gravity into question. A modern thinker might wonder why ideas built around an unmoving Earth at the center of the universe were not tested sooner. But this was a way of thinking that was alien to the Greeks. The primary proof of a theory was through argument.

Famously, Aristotle stated that men and women had different numbers of teeth. He never bothered to check this, nor did his contemporaries, as experiment and observation were not the regular means of performing natural philosophy, the predecessor of science. Instead, different individuals would put forward opposing theories and have a debate to discover which theory was most effective—an approach we still apply in the law courts. The senses, it was argued, could always let you down. It was better to stick to logic and argument.

It is entirely possible that some other theory could have taken hold with equal strength. The four elements model of matter was challenged by an idea that everything was made up of atoms, which were the uncuttable (*atomos*) limit of slicing an item in half repeatedly. But that idea lost the philosophical debate. There could have been other models that had captured the imagination of those ancient philosophers equally well.

An amateur theorist, for example, has recently suggested that gravity doesn't exist and instead all matter is expanding.[5] We wouldn't notice this, it is claimed, because the matter we are made up of is expanding, too. So what we see as objects falling to the ground is in fact the expanding ground meeting up with the expanding objects. It's a theory that could have flourished readily in the armchair debating environment of ancient Greek science, though it is now easy to find tests that disprove it.

Probably the clearest picture of the Greek model of gravity comes from Aristotle's *The Physics*. This book influenced thought on many aspects of the physical world for 2,000 years, despite being almost entirely incorrect from end to end. Here we meet Empedocles' four elements, plus that fifth heavenly substance. Here also is the idea of different elements having different tendencies to move, not through any force being applied, but simply due to their nature.

This movement has to be understood in the context of totally non-Newtonian framework. Aristotle did not see motion as the outcome of an applied force, but rather as the realization of something that has the potential to be. (If that sounds vague, it is—we are dealing with philosophy, after all.)

In Aristotle's universe an object is naturally at rest. If you push an object, it moves, but stop pushing and the motion will cease of its own accord. The place where it stops moving depends on its elementary nature—so Earth, for example, stops moving when it is as close as it can get to the center of the universe. This is nothing to do with concepts like force and energy. It was thought that it was the nature of these elements to behave in this way just as it was in the nature of a dog to chase a cat.

In the world of Aristotle and his contemporaries, earthy objects had a natural heaviness (gravity) that made them want to be at the center of the universe, while airy objects had natural lightness (levity) that made them move away from the center. The amount of the appropriate element that was thought to be in an object influenced how fast it accelerated toward its preferred resting place. With lots of earth, and hence heaviness, an object would tend to move faster toward the center than one with more of a balance, because that urge to reach its natural place was stronger.

This was a simple enough hypothesis to check, had Aristotle tolerated any idea of testing theories. Does a big rock fall much faster than a small rock? Aristotle's theory required this, because the big rock had much more of the earthly element in it, all with a strong desire to head toward the center of the universe. But there seems to have been near-total blindness to any evidence from experiment.

Aristotle was not saying that earthy substances had more atoms in them, or were made up of heavier atoms to give them this tendency. It always came back to the nature of the objects.[6] Just as, for example, a chair had the nature of being something to sit on where a needle didn't, so different types of matter had their own natural tendencies. Just as an animal naturally grows and reproduces, so the different types of matter behaved with gravity or levity. It wasn't mechanics, it was *nature*.

When an object was thrown, this was an unnatural action, working against the object's nature, something Aristotle called a violent movement. An object could only move away from its natural tendency because an external cause influenced it. Aristotle had no truck with "action at a distance"; for him, if something was made to move, the cause had to be in contact with it. This provided a bit of a challenge when thinking about a thrown rock. It was fine that it started moving, as the hand was touching it. But according to this theory, as soon as the rock left the hand, natural motion should take over and it should plummet to the ground.

Even Aristotle's disdain for checking theory against observation couldn't miss the fact that a hurled rock doesn't immediately stop moving forward and drop to the Earth as soon as it leaves the thrower's hand. To get around this problem, he suggested that the initial push was given to both the rock and the surrounding air. And it was the air, which remained in contact with the rock, that kept it going until its natural motion toward the center of the universe took over.[7]

Did this picture of rock and air interacting reflect what was actually seen to happen? Once more we have to remember that using experiment to test theory was not considered an essential by the ancient Greeks. The senses could be fooled. It was better to rely on logical argument. This Greek legacy and view of gravity would continue to hold sway all the way to Galileo. But it did not go entirely unchallenged.

CHAPTER THREE

GRAVITY FALTERS

> *Physicists consider that the motion of heavy bodies downwards is altogether natural, and the motion of light bodies upward is likewise wholly natural, so that as a result they suffer no strain.*
>
> —*Opus Majus* (1267)
>
> Roger Bacon

If you believe many histories of science, the world went into an intellectual coma between the ancient Greeks and the Renaissance. In this picture, everyone suddenly stopped being interested in the world around them and totally ignored it for around 2,000 years, relying on whatever the Greeks had said as truth and getting on with their lives. They just didn't seem to care about the natural world and how it worked.

This isn't a totally fictional idea. There certainly was less original scientific thought in this complex period of history than took place before or since. But equally it wasn't an intellectual desert. Some have suggested that the rise of first Christianity and then Islam in this period resulted in a suppression of interest in nature—

but that, too, is a lazy analysis based more on prejudice than the historical record.

At all times in their existence, these religions have supported the idea that to study nature is to study the handiwork of God, and so it is a good thing. There have been times, it's true, when both religions have had factions that opposed intellectual thought and disliked specific scientific theories, particularly when they appeared to run counter to scriptures. But neither faith made it impossible to come up with scientific ideas in the dark ages and medieval periods—in fact most of the scientific work done then had explicit support from the religious hierarchy.

There was, without doubt, a lot of emphasis put on the words of the Greek authorities, and Aristotle in particular was regarded as the best source for much knowledge. Yet medieval natural philosophers were sometimes prepared to challenge the ancient Greek ideas with thoughts that were inspired more by observation and less by pure argument. Even Aristotle was occasionally questioned.

Two scientific movements emerged from those apparently suppressive religious cultures. First, between the seventh and thirteenth centuries AD, the Islamic world took ancient Greek natural philosophy, threw out much of the garbage, and added many of their own insights. Then, as Islamic science started to decline, the Christian West picked up the baton, learning much from their Middle Eastern neighbors and making their own developments that would eventually lead to the flowering of the Renaissance.

Both of these medieval bursts of natural philosophy would

contribute to transforming the dead-end concepts of the ancient Greeks into the foundations that made first Galileo's and then Newton's work possible. Two specific aspects of the Greek view of science began to be eroded in the medieval period. The first we have already met—the reliance on logic and reasoning rather than direct observation of the external world. The second was a move from Aristotle's "science" based on qualitative concepts to one that relied on number and measurement.

There had, of course, been Greek philosophers who understood the importance of mathematics. Pythagoras believed that the whole universe was based on number, while Euclid pulled together a powerful geometrical toolkit and Archimedes made considerable use of number and geometry in his science and engineering. But Aristotle presented the medieval natural philosophers with a worldview that kept *number* less important than *qualitative* measures and this approach proved hard to overcome, in part because of strong religious support.

Even so, the increasing use of technology, pragmatically requiring a number-based approach, and the importation of the powerful Indian number system weakened Aristotle's dominance. At the same time, the rediscovery of Greek mathematical texts and the new mathematical wisdom coming from Arabic mathematicians meant that there was a new enthusiasm for quantitative methods, coming closer to the precision we now associate with science.[1]

At the same time, the Greek disdain for observation as a basis for natural science was being eroded. Roger Bacon, the thirteenth-century scientist friar, was one of the first to explicitly emphasize the importance of experiment. Bacon is a fascinating character who became the center of mystical mythology. By Shakespeare's time he was represented as a dark philosopher magician who built a talking brass head that could answer deep questions and who constructed a magical wall around the whole of Britain to prevent invasion by the perfidious French (an idea that would have been much appreciated in Tudor times).

The real Roger Bacon had a story far removed from magic—in fact Bacon explicitly ridicules magic as a tool used by fraudsters to extort money out of gullible marks—but his actual life was in some ways more extraordinary than the legend.[2] He had the misfortune to be a man who was obsessed with collecting and passing on scientific knowledge at a time when the head of his Order (the relatively newly formed Franciscans) had decided that the friars had become too academic and should get back to their original mission of helping the poor. As part of this move, Franciscans were banned from writing books.

Bacon was consumed by the idea of producing an encyclopedia of science. To get around the ban on book writing, he gained the support of an influential cardinal who agreed to back Bacon's project as long as he kept things secret—including the cardinal's involvement. This seemed a dangerous basis to proceed. Bacon could well have given up at this point had he not received startling news. In an unexpected ballot result, that same supportive cardinal was elected pope.

This seemed a literal godsend. Bacon started to write a proposal for the new pope to explain what would go into his encyclopedia. The proposal grew and grew until it contained over 600,000 words. It was a proposal 8 times the length of the book you are currently reading. Bacon realized he needed a covering letter for such a weighty tome. But there were things he had missed out of the original proposal, and new information on nature seemed to reach him every day. The covering letter, too, became a sizeable volume in its own right.

Unbelievably this happened again with his next attempt at a letter, so when Bacon finally had produced a simple letter to the pope explaining his project, it accompanied three volumes totaling over a million words. This was sent off with a trusted manservant to Rome. At the time the journey from England (Bacon was based in Oxford) was not trivial and it was some weeks before the heavy manuscripts arrived. Bacon's man John must have been baffled by the upheaval that greeted him in the heavenly city. The place was in chaos. The reason? The pope had died while Bacon's masterpiece was on its journey.

Bacon must have been horrified when he learned of the death, all the more so because the new pope turned out to be a hardliner who had no interest in science. What happened next is not certain because the only information we have comes from a book that was written 100 years later. Bacon is said to have been thrown into jail for many years until a more understanding figure took the papal throne. By then the friar was in his seventies and had no stomach for the project. He would write again, but only on theology.

Although the great encyclopedia was never produced, Bacon's

massive proposal has survived. To all intents and purposes a summary of scientific knowledge of the day, it includes a good number of original ideas from Bacon himself. We see some interesting concepts developing, though he clearly finds it very difficult to part from Aristotle. But what is particularly important is what Bacon had to say about experiment.

We tend to think of the modern scientific approach developing in the interplay between Galileo's early experimental science and the organizing thrust of the politician philosopher Francis Bacon (no relation to Roger as far as we are aware) in the seventeenth century. But in the first volume of Roger Bacon's proposal, known as the *Opus Majus,* its author provides a whole section dedicated to the importance of experiment. Bacon says:[3]

> He therefore who wishes to rejoice without doubt in regard to the truths underlying phenomena must know how to devote himself to experiment.

We have to be a little careful about exactly what he meant by this. The original was written in Latin and the word being translated as "experiment" could just as easily mean "experience." Although Bacon does document actual experiments, some of which he appears to have carried out himself, many of his observations on nature are more like travelers' tales heard in the tavern—sources that would not be given scientific credibility today.

Yet even so, we should not underestimate how fundamental a shift this was from the ancient Greek ideas to which Bacon still so carefully gave lip service. He was elevating observation and experiment above logic and argument, turning the accepted order of science and discovery on its head.

For someone who seemed to have a lot of time for Aristotle, Bacon also went against the traditional Greek view on the lack of importance of a quantitative approach to science. For him, mathematics was crucial:[4]

> Of these sciences the gate and the key is mathematics . . . he who is ignorant of [mathematics] cannot know the other sciences nor the affairs of the world, as I shall prove. And what is worse men ignorant of this do not perceive their own ignorance, and therefore do not seek a remedy.

By Bacon's time, one of the ancient Greek ideas that was causing real concern was the description of how the planets moved and what kept them traveling on their journeys through the heavens. The accepted view was that they were fixed in crystal spheres, powered by some undefined external force, the prime mover, a phenomenon that was often identified with a deity taking the part of the organ grinder at the handle of a celestial barrel organ.

Medieval philosophers were concerned that the old concept of crystal spheres was no longer sustainable. The development of the planetary motions called epicycles—designed to account for the odd observed behavior of the planets that had some of them reversing in their tracks—required the planets to move along circles within circles, somehow breaking away from the spheres.

The popular theory of the time did away with a prime mover, something that turned the whole celestial mechanism with a single action. Instead, it was suggested that the movement of each of the spheres and sub-spheres in the epicycles should be assigned to angels.[5] It's hard when hearing this term not to conjure up a picture of creatures with white feathered wings (and quite possibly harps), fluttering around pushing the spheres, but it seems to have been more a metaphorical use of the term "angels" to represent some kind of impersonal heavenly forces.

Just as many have assumed incorrectly that the medieval thinkers blindly thought that the Earth was flat, there is also the assumption that until Copernicus came along, they were all absolutely certain that the Earth was fixed and unmoving at the center of the universe. Although most philosophers of the day did leave the Earth in place, because of difficulties with operating Aristotle's model of gravity if the Earth wasn't central, a number of medieval thinkers suggested that the Earth was rotating.

If the Earth rotates it makes the way that heavenly bodies apparently dash around at great speed much more acceptable. Specifically, it seemed much more likely that the relatively small Earth should rotate once in 24 hours than that the whole vast panoply of the universe should flash through the heavens in this time.[6] At a stroke, by making the Earth rotate it was possible to remove the majority of the movements in the heavens, leaving only bodies like the Moon, planets, and comets to explain.

Working in the early years of the fourteenth century, John Buridan and his younger assistant Nicole Oresme argued for such a rotating Earth. In their minds it even fitted better with Aristotle's design for the universe. After all, he had said that everything below the Moon was corrupt and subject to change, while everything above its orbit was pure and unchanging. Transferring movement from the heavens to the Earth made sense in this picture, because it stopped all the regular daily variation above the Moon.

However the basic gravitational principle that heavy objects had a natural tendency to fall to the center of the universe (hence putting the Earth at the center of everything) was less likely to be challenged. Here is the twelfth-century scientist/monk Adelard of Bath, in an explanation that shows just how alien the viewpoint of the time, based more on the behavior of human beings than inanimate objects, now seems:[7]

> What is heavy stays best in the lowest position. Every thing loves that which preserves its life. But it tends towards that which it loves. Therefore it is necessary that every earthy thing tends towards the lowest of all positions . . . Where individual weights hurry to, towards that place they also fall.

Although Aristotle's approach to gravity would largely hold until the Renaissance, it was questioned by a number of the great Arab scientists, who seemed to feel that there was something closer to an attraction taking place than a natural tendency—an idea much closer to our modern image of gravity. This change in view may have been helped by the increasing awareness of magnetism.

Roger Bacon appears to have been an enthusiastic follower of a French philosopher Peter of Maricourt, often referred to as Peter Peregrinus, probably because he had been on a lengthy journey (a peregrination)—perhaps a pilgrimage or a crusade. Peter wrote a treatise called *De magnete,* which provided the first proper study of the effects of magnets.[8] It was very popular in medieval times and could well have been part of that spark of interest in the possibility that gravity was the result of a similar, "action at a distance" type attraction.

Part of the impetus to see gravity in this way came from outside the influence of the heavy hand of the ancient Greeks. By this time, ideas were filtering in, along with the vital number system, from the mathematical hotbed of India. So, for example, the Arab philosopher al-Biruni was influenced by the seventh-century Indian mathematician Brahmagupta who explicitly states that the objects fall to the Earth as a result of attraction.

Somehow, though, despite its many failings, Aristotle's picture of gravity as a natural tendency remained the dominant understanding. The ancient Greek worldview was too strongly embedded in the approach taken by medieval scholars. It would take an iconoclast from Italy to shake things up in a way that could never see a return to this obscure mode of thought, and an irritable English eccentric to transform gravitation into something we would recognize today.

CHAPTER FOUR

AND YET IT MOVES

> *And thus Nature will be very comfortable to her self and very simple, performing all the great Motions of the heavenly Bodies by the Attraction of Gravity which intercedes those Bodies, and almost all the small ones of their Particles by some other attractive and repelling Powers which intercede the Particles.*
>
> —*Opticks, Second Edition* (1718)
>
> Isaac Newton

The most dramatic step in the movement of science away from the ancient Greek position was replacing the Earth with the Sun at the center of the universe. This change in viewpoint might not seem to have much influence on our understanding of gravity, but by removing the Earth from the center of everything it invalidated the basic logic that heavy things have a desire to move toward the universal hub.

Although it seems obvious to us today, because we are so used to it, the idea that the Earth moves around the Sun isn't a particularly common sense assumption. It appears quite clear that the Sun moves through the sky, passing around the Earth during the day. We still talk of sunset, not horizonrise. But all the way back to the ancient Greeks there has been an awareness that movement is relative, making it difficult to say what is moving and what is stationary.

Most of us have experienced that strange feeling, sitting on a train in a station, when an adjacent train starts to move. It can easily seem for a moment as if our train is moving in the opposite direction. (Unnerving if the train is in a terminus.) We actually feel movement until our brains untangle what is really happening. With no other clues, it is impossible from sight alone to work out if an object we are on is moving with respect to its surroundings, or whether the surroundings are moving with respect to it.

The same applies to the Sun as seen from the Earth. In principle, the Sun could be moving around the Earth or the Earth around the Sun and there are few direct confirmations of which really is the case. We know that one ancient Greek astronomer had already realized this in the third century BC. We don't have the book in which Aristarchus of Samos puts forward this theory, but his younger contemporary Archimedes makes a tantalizing reference to the idea.

Archimedes had written a book called *The Sand Reckoner,* in which he calculates how many grains of sand it would take to fill the universe.[1] This may seem a totally futile activity, on a par with working out how many angels can dance on the head of a pin, but Archimedes was a man on a mission. He was frustrated by the limitation of the Greek number system and wanted to demonstrate to his patron, King Gelon of Syracuse, that the approach then used could be bettered.

The biggest number in the Greek system was a myriad—10,000. You could go as far as 100 million by saying "a myriad myriads" but that was the limit of numbers, and was pathetically small when dealing with a challenge like counting the grains of sand on a beach, let alone the number that would fill the universe. So Archimedes produced a whole new system that would work indefinitely, starting at that "myriad myriads" level and working upward. To demonstrate the power of his idea, he calculated the number of grains of sand it would take to fill everything.

The understanding of the universe at the time was roughly what we would call the solar system—the Earth, the Sun and Moon, the planets, and the stars, which were thought to be fixed to a sphere just outside the orbit of Saturn. This all rotated around the Earth at the center of everything. However, to cover all bases, Archimedes also did a calculation based on a different universe:[2]

> Aristarchus of Samos brought out a book consisting of some hypotheses, in which the premises lead to the result that the universe is many times greater than that now so called. His hypotheses are that the fixed stars and the sun remain unmoved [and] that the earth revolves about the sun in the circumference of a circle, the sun lying at the middle of the orbit.

Archimedes was frankly rather doubtful about Aristarchus's reasoning, pointing out how few supporters it had, but he took a look at it and decided that a heliocentric system would be a significantly bigger universe than the conventional one (though still pretty much on the scale of the solar system). He never took the idea seriously—it just made for an interesting exercise in mathematics. And most other natural philosophers would simply ignore the concept of shifting the Earth from the center of the universe until Nicolaus Copernicus came along.

Copernicus (more properly Niclas Kopernik, as we know him by the academic Latin version of his name) was born in Thorn (now Torun), Poland in 1473. Though young Copernicus studied liberal arts in Krakow and then canon law at the University of Bologna, he lodged with a math professor who encouraged an interest in astronomy. But this was very much a hobby. His next educational stopover was to study medicine at Padua, before finally taking a doctorate in canon law.[3]

Back in Poland, while technically a canon of the chapter of Frombork, Copernicus split his time between helping administer the diocese of his uncle Bishop Lukasz and practicing as a medical doctor. But in his spare time he was still fascinated by astronomy, which lead him to write two books on the motion of the heavens. The first, *De Hypothesibus Motuum Coelestium a se Constitutis Commentariolus,* seemed to have passed by pretty much unnoticed, but the second would cause a scientific revolution.

Written about 15 years later than his first title, his second book stayed on the shelf for 13 years before being published. Called *De Revolutionibus Orbium Coelestium* it firmly placed the Sun at the center of the universe with the occupants of the heavens (including the Earth) rotating around it in orderly fashion. It is hard to believe it was a total coincidence that the book was printed days before his death on May 24, 1543. Copernicus knew this was a book that would cause strife.

There is no doubt that all the time and effort that Copernicus put into making astronomical observations helped him come to his universe-reshaping conclusion. He spent many of his evenings, in the clear darkness that we would never experience in a city now, haunting the tallest towers he could gain access to, far above the street life, studying the movement of the planets using wooden and metal instruments without lenses.

Everyone knows that the Catholic Church at the time fiercely opposed the idea that the Sun was at the center of the universe. At least, that's the simple version of the story. In reality the situation was more finely nuanced. It's true that the Church had objections to the Sun being fixed. There are two references to the Sun stopping in the sky in the Bible, which seems to imply that it moves. And the Church revered Aristotle as an authority. Yet it was prepared to modify Aristotle's ideas where necessary. (He had said, for example, that the universe was eternal, which was conveniently ignored to match scripture.)

Bear in mind also that Copernicus was an active member of the Church hierarchy, not some troublesome outsider. In fact *De Revolutionibus* was written with the encouragement of the Catholic Church and dedicated to the pope. Yet there were those around Copernicus who were uncomfortable with the book and urged caution, which may explain why its publication was held back.

The book's editor replaced the original introduction, apparently without Copernicus's consent, with a warning that these were just ideas and hypotheses that shouldn't be taken too seriously. And before long there would be outright condemnation of the entire work.

It seemed at the time that all efforts were focused on the implications that Copernicus's idea had for the orbits of the planets. His model simplified things a great deal. By making the Earth rotate around the Sun between Venus and Mars, all the strange behavior of the planets, including the need for them to reverse in their tracks, was removed. They could now travel on what were assumed to be neat, perfectly circular orbits. Yet there was another implication if Copernicus was right. The concept of gravity that had been held since Aristotle's time would have to be thrown away.

The old model of the universe might have produced some weird orbits, but it made sense for things to fall toward the Earth if that was the same as the center of the universe. But if the Sun were now the center, with the Earth demoted to a mere orbiting planet, there had to be a different explanation for gravitation. After all, things didn't tend to fly toward the Sun when you let them go, they fell to Earth. Rather than a natural tendency toward the center, this seemed more like an attraction between bodies.

Two important figures followed Copernicus in our understanding of the solar system. Tycho Brahe, born in Denmark just a few years after Copernicus died, would contribute many detailed observations of the movements of the stars and planets from his observatory at the romantic-sounding Uraniborg, a purpose-built observatory on Hven island that lay between Zealand and Scania.

Brahe was certainly a character. He famously had the bridge of his nose cut off in a duel, and for the rest of his life wore various metal prosthetics, giving him the appearance of a medieval cyborg. He supported the astronomical benefits of the Copernicus arrangement, but came up with a compromise that would keep the Earth at the center of the universe (leaving gravity to act "properly") while still explaining the odd movements of the planets.

Tycho Brahe suggested that the Moon and the Sun revolved around the Earth, while all the other planets traveled around the Sun. This might work in terms of simple correspondence to what is observed, being the "correct" relativistic picture from the viewpoint of the surface of the Earth, but it now seems a very artificial construct, giving the Earth an unnaturally different basis to all its sister planets.

That Brahe's model went against nature was certainly the view of the other man to contribute specifically to our picture of the workings of the universe. This was German astronomer Johannes Kepler. Twenty-five years younger than Brahe, he tried unsuccessfully to get the Dane to see that having the Sun at the center of things made more sense. Kepler had his own strange beliefs, mostly about a mystical geometric structure for the universe that put each of the planets in an orbit whose size was determined by one of the Platonic solids (tetrahedron, cube, octahedron, etc.—though not in this order). But he was a strong supporter of Copernicus.

Kepler used Brahe's comprehensive data to develop some remarkably prescient ideas on the way the planets moved, realizing that they traveled in ellipses rather than the circles that had been assumed since ancient Greek times, and producing laws that described the speed with which planets moved and the time it took them to travel around their orbits.

Although these laws have implications for the effects of gravity, as Newton would discover some time later (see page 58), Kepler did not seem to have given the mechanism by which planets stayed in their orbits much consideration. But this could hardly be said of the man who became famous for risking his life to support the heliocentric model, the great Galileo Galilei.

Galileo was, in fact, 7 years older than Kepler, born in Pisa in Italy in 1564.[4] His predecessors were very much *academic* rebels. Copernicus, for example, could see that there was something wrong with the traditional view that put the Earth at the center of the universe, but could only present his argument in an intellectual fashion. Galileo had something more. He was, without doubt, a scientific genius, but he was also a slick publicist. He was a rebel more of the gut than the intellect. He had a deep down feeling that the dependence of science on Aristotle's theories was plain wrong, and he wasn't about to give in merely to conform.

This rebelliousness, along with an interest in mathematics and the intellectual strength to challenge received wisdom, seems to have been inherited from his father, Vincenzio. The older Galilei was a musician who liked to take on the received wisdom of the day, saying in a book on music that those who rely on authority "act very absurdly." Despite his enthusiasm for mathematics, though, he was none too pleased when his son gave up a medical degree to concentrate on math and science.

The rebellious streak was stronger in Galileo than it had been in his father, and there was no stopping him. During his time at university, he had been bombarded with writings that paid absolute respect to the theories of the ancient Greeks, and in particular Aristotle. For Galileo ideas were pointless without experiment and observation to back them up. He felt it was necessary to strip back the received wisdom and build a new science based on what was actually seen to happen.

One of the best-known stories about Galileo seems to illustrate this attitude. Aristotle had said that different objects fell to Earth at different speeds. The heavier they were, the greater the urgency with which they sought out their natural position. The more weight they had, the faster they scurried toward the center of the universe. Aristotle would never have thought of actually trying this out, but Galileo was made of different stuff.

According to legend, the Pisan Galileo dropped balls of different weights off the crazily leaning bell tower of his home city's basilica and observed that they fell at the same rate, proving Aristotle wrong about this fundamental behavior of gravity. Sadly, there is a big problem with this story. Galileo was never one to keep his successes quiet, yet he never mentioned undertaking this experiment. We hear of it when an assistant was pulling together stories of Galileo near the great man's death. It is more than likely that it never happened.

This doesn't mean that Galileo wouldn't turn the Greek ideas of gravity on their heads, but initially his experiments in natural science had to take second place to earning a living. When he failed to get his medical degree he missed out on a steady income and was forced instead to make money however he could—inventing, teaching, doing whatever would bring in some cash. This proved doubly necessary when his father died and Galileo became responsible for his sisters' dowries.

The stability that Galileo craved to get on with his work was mostly provided by positions in Italian courts, meaning that he was always looking for a new opportunity to dedicate some piece of work to the latest nobleman with whom he hoped to win favors. This didn't stop Galileo from doing some important work—but it seems typical of his character that his first great breakthrough should be one where he had his eye to the main chance and which involved a certain amount of skullduggery.

It is often said that Galileo invented the telescope. He didn't. By Galileo's day, telescopes were already being made elsewhere in Europe, particularly by the spectacle-makers of Holland, and Galileo heard that a Dutch optician was heading to Venice to try to interest the local nobility in this new device. Luck was on Galileo's side. The Venetian court asked a friend of Galileo, Friar Paolo Sarpi, to investigate the new technology. Sarpi kept his Dutch visitor busy while Galileo hurriedly assembled his own telescope and hurried to Venice to show it off.

It was a huge success. The elderly Venetian noblemen had to be restrained from getting into an unseemly tussle over who would be allowed to take the telescope up on the roof next to scan the ships in the lagoon. To the leaders of a seafaring city state, the benefits of the telescope were very obvious. The temptation must have been great to sell this new toy to the highest bidder to raise some much-needed cash, but instead Galileo showed political wisdom in presenting his telescope as a gift to the doge of Venice, in return for which he was offered a lifetime position at the University of Padua on double pay.

The telescope would do more than improve Galileo's finances. With his later models he began to unpick Aristotle's outdated view of the universe with a number of observations that simply couldn't be squeezed into the old picture. Galileo studied the Moon, producing beautiful drawings of the mountains and craters on the surface, details that were impossible to distinguish with the naked eye. According to Aristotle, the Moon was a perfect sphere—to Galileo it was increasingly obvious this was not the case.

Soon after, he made a much more dramatic discovery, something that would have been impossible without the telescope. Studying the planet Jupiter he noticed four points of light that moved around the planet in a way that over time could only be explained if they were moons, circling the distant body. Galileo's political awareness came to the fore again as he named this new discovery the Medician Stars after the powerful duke Cosimo de' Medici. In return, he got a new post at Pisa with the same pay as his old post, but with no teaching duties and with money up front. He had yet to see a penny from his position in Padua.

The discovery of Jupiter's moons was the final straw for Galileo when it came to moving the Earth from the center of the universe. A fundamental tenet of Aristotle's worldview was that everything had to rotate around the Earth. Yet Jupiter's moons clearly rotated around that planet, not the Earth. Once a single body disobeyed the rules, Aristotle's whole model fell apart. It was time, Galileo felt, to put the Copernican universe in its proper place.

What happened next is often represented in black and white, as if it were an old-fashioned Western movie. Brave Galileo opposes the Aristotelian universe and is cruelly punished by the evil Catholic Church. Good guy versus bad guys. Without a doubt Galileo was badly treated, but he showed an unusual lack of political awareness in the way he went about publishing his ideas, and suffered as much as a result of this poor judgment as for his original thinking.

He structured his book on the new model of the solar system, called *Dialogue Concerning the Two Chief World Systems,* first published in 1632, in the traditional form of a discussion. This was classical combative philosophy in written form. The two participants put forward opposing views—one, the universe of Aristotle and his supporter the Greek astronomer Ptolemy; the other, the Copernican model. The book takes the form of a discussion between the two, a surprisingly ancient Greek approach to one of the first great challenges to classical thought.

The book was not sneaked out and published in secret, hidden away from the ecclesiastical hierarchy. Galileo submitted his work to the official censor and made all the changes required to keep the Church happy with the contents. However, he indulged in a spot of fun that was to have dire consequences. The character supporting the ancient Greek view was called Simplicio. There was a Greek philosopher of this name, but it's hard not to see Simplicio as the simpleton, putting forward the idiot's view. Hardly a balanced picture.

To make matters worse, one of the requirements of the censor was to put in a postscript emphasizing that the Church supported the Earth-centered universe, and that the Copernican view was merely an intellectual exercise, and could not be taken as a meaningful description of reality. Galileo did include this postscript, but the words, arguably the words of the pope himself who had clearly expressed this view, were not left as neutral comment, but put into the mouth of Simplicio. Was Galileo calling the pope a simpleton? He had made some enemies along the way and now he had given them perfect ammunition. He was charged with heresy.

A fair trial could still have gone Galileo's way, but the Inquisition had a way of getting the results they wanted and he was found guilty on what was little more than a technicality. Before sentencing, Galileo swallowed his pride and admitted that he had been mistaken. It is possible that he could have been burned at the stake for such a crime, but instead he was committed to life imprisonment.

It's at this point that another legendary aspect of Galileo's life is supposed to have taken place, and just like the balls of the leaning Tower of Pisa, the chances are that this legend is untrue. After his retraction, he is supposed to have muttered "*Eppur si muove*"—"and yet it moves"—defiantly supporting the idea that the Sun is not fixed in place. Under the circumstances, this is very unlikely. The Inquisition was not a body that took rebellion lightly, and there was no evidence at the time that the sub voce mutiny took place, even in the biography that produced the Tower of Pisa legend. It seems to have been added to the Galilean mythos about 100 years later.

Thankfully, Galileo's life imprisonment ended up in the form of a relaxed house arrest, leaving him able to continue his work. And it was while serving his sentence that he wrote the book that would build on the Copernican challenge to the old idea of gravity to provide a very different way of looking at the movement of bodies. This book was to be his masterpiece of physics, *Discorsi e dimostrazioni matematiche, intorno à due nuove scienze* (*Dialogues and Mathematical Demonstrations Concerning Two New Sciences*).

Like its more famous predecessor, the book was in the form of a discussion, here between three people—Salviati, who represented Galileo's viewpoint, Simplicio, who was once more stuck in the mindset of ancient Greece, and Sagredo who acted as an independent observer and referee of arguments. With his masterpiece assembled, the old Galileo would have rushed into publication, making sure that a rich patron's name was splashed widely across the front. As it was, he now found himself in a different world.

The Italian publishing houses would not touch anything he wrote—he was damaged goods. He toyed with the possibility of publishing in Germany, in Vienna, in Prague, yet even this far from Rome there was a danger of a backlash from the Inquisition. In the end, it was the visit of Dutch publisher Louis (Lodewijk) Elzevir to Italy in 1636 that seemed to open an ideal pathway to a safe, Protestant printing of his work. Elzevir, the third Louis in that great publishing family, took an incomplete manuscript back to Leyden.

The final book, printed in 1638 was not quite what Galileo had hoped for. As is so often the case, for example, the author felt irritated that the publishers had changed his title.[5] We don't know what Galileo had intended it to be, but he called the printed title—in full translated as *Discourses and Mathematical Demonstrations concerning Two New Sciences pertaining to Mechanics and Local Motions*—that the publishers had substituted for his own, "a low and common title for the noble and dignified one carried upon the title page."

Despite publishing in Holland, Galileo was aware that the book would not go unnoticed in Italy. He tried to get around this by expressing his deep surprise that the book had been published at all. He did this in the book's dedication to a long-time supporter, the Count of Noailles. He says that he just sent out a few copies for people to look at, and the next thing he knew, Elzevir was about to publish it and asking for a dedication.[6] Whether or not the Inquisition was taken in by this naïve attempt to cover his tracks, there was no reprisal from the authorities.

Galileo's book is divided into 4 "days" of discussion, and in Day 1 he covers pendulums, which rely on gravitation for their motive power, and that inevitably inspires thought about the force of gravity. Legend has it that long before his trial, Galileo was in the cathedral in Pisa, and (rather bored during a sermon) started watching a candelabrum, suspended from the ceiling on a chain. It had been pulled to one side so the candles could be lit and after being released had yet to settle down—it was swinging from side to side.

According to the story (almost certainly as fictional as the other legendary moments in his life) what Galileo noticed, to his surprise, was that that though the swing of the candelabrum on the end of the chain was getting shorter and shorter as it "ran down," the length of time it took to make the swing remained the same.

Whether or not he was inspired by seeing a swinging chandelier, Galileo certainly undertook a series of experiments on simple pendulums made with stones tied to the ends of pieces of string. Although the period of the swing varied, as you might expect, with the length of the pendulum, neither the size of the swing nor, crucially, the amount of weight attached to the string changed the time it took to go through its regular motions.

Ever since Aristotle's time, people had been convinced that heavy bodies fall faster than light bodies. In the end, a pendulum was nothing more than a falling object, deflected off its path by a piece of string. The stone on the end of the string was falling, if in a constrained fashion. Surely if heavier objects fell faster, as Aristotle claimed, then a pendulum with a heavier weight on the end should swing faster?

Galileo had already torn into Aristotle on this subject before getting on to pendulums. "I greatly doubt," he says, "that Aristotle ever tested by experiment whether it be true that two stones, one weighing ten times as much as the other, if allowed to fall at the same instant from a height of, say, 100 cubits, would so differ in speed that when the heavier reached the ground, the other would not have fallen more than 10 cubits."

Simplicio, the character in Galileo's book who supports Aristotle tries to defend his hero, saying that it sounds as if Aristotle did the experiment because Aristotle says "we see the heavier . . ." which implies something to see. But Sagredo, speaking for Galileo, will have none of it, because Galileo really *had* done the experiment.[7] According to Galileo, dropping a cannon ball weighing 1 or 200 pounds only reached the ground "a span ahead" of a small musket ball, weighing only half a pound.

What he doesn't say is that it is very difficult to drop two balls like this and be sure of the timing of their fall, hence that observed gap between the two balls. (Though it is obvious he was aware of this because when he goes on to make experiments with much more precise timings he used a totally different approach.) He does, however, point out that varying air resistance can have a significant effect on the rate of fall.

Galileo goes on to demolish Aristotle's picture through a

clever piece of logic. Imagine you had two balls with different weights that *did* fall at different speeds. Then imagine what would happen if you tied a piece of string between them. The slow one should slow down the fast one, and the fast one should speed up the slow one—together they should fall at an intermediate speed. Yet by fastening them together you are making a single object with an even greater weight than the heavier of the two, so the combined object should fall faster than either of the individual objects. This logical flaw makes it sensible to infer that objects with different weights must fall at the same speed.

On August 2, 1971, the true extent of Galileo's assertion was demonstrated graphically by *Apollo 15* mission commander David R. Scott. During the third moonwalk of the mission, Scott took a geological hammer and a feather out on the surface of the Moon. Immortalized on video, he can be seen dropping the two items together. It's not an experiment Galileo could do because of the impact of air resistance. But with no atmosphere, the hammer and the feather fell on the Moon at the same rate.[8]

In his book, Galileo suggests we move from dropping things to observing pendulums, in order to make the experiment more controlled. This is long after the fabled observation in the cathedral, but there seems no doubt that Galileo had spent time experimenting with different pendulums. He describes letting two pendulums go simultaneously, one with a cork ball on the end, the other with a lead ball, "more than a hundred times heavier" suspended from it.

With equal length strings, even after hundreds of swings, the two pendulums kept step—the very different weights were falling at the same rate. He goes on to point out that this was even true when the cork weight, which had a swing that reduced in size much quicker than the lead, was only moving a fraction of the distance the lead was.[9]

Funnily enough, this observation of Galileo's about the frequency staying the same however far the pendulum is swinging is, in fact, wrong, despite often being repeated.[10] In a conventional pendulum it only holds true for a small swing; once the pendulum moves more than about 15 degrees either side of the vertical, the forces acting on it cease being linear, no longer producing a consistent frequency for any swing size. It is very likely that Galileo was aware of this, but he seemed happy to brush over it and ignore it to make his more important point.

Pendulums would continue to play an important role in the study of gravity. In fact Newton's contemporary Robert Hooke suggested in 1666 a way to use a pendulum to measure the acceleration due to gravity.[11] By taking a standard pendulum to different points on the Earth's surface, Hooke suggested, it would be possible to check out relative variations in gravity, and by implication variations in the radius of the Earth. This approach would continue to be used as late as the 1930s. In 1936, for example, the U.S. National Bureau of Standards used a pendulum to determine the gravitational acceleration in Washington, D.C., to be 980.080 ± 0.003 cm/s^2.

The most direct speculations about gravity in Galileo's book, though, occur in Day 3, which starts out with a statement outside of the conversational format, where the reader is addressed directly by Galileo. "My purpose," he says "is to set forth a very new science dealing with a very ancient subject."[12] It is about movement—change of position—and specifically includes acceleration due to gravity. After dealing with uniform motion, he gets on to "naturally accelerated motion." (This is as opposed to "violent motions," which are those where there has been an intervention, like giving something a push, a distinction that Galileo maintained from the ancient Greeks.)

I won't dwell too much on the detail of Galileo's arguments, much of which are bogged down in geometric descriptions of different situations, but there is a crucial line in his introduction to his work on accelerated motion. He says that he is attempting to make his definition of accelerated motion "exhibit the essential features of observed accelerated motions" and that "In this belief we are confirmed mainly by the consideration that experimental results are seen to agree with and exactly correspond with those properties which have been, one after another, demonstrated by us."[13]

Although this is painfully wordy, it is a crucial statement. Here is Galileo throwing out the ancient Greek idea that it is enough to come up with a good theory, debate it, and the "best" theory becomes accepted wisdom. Instead we have the simple foundations of the modern scientific method. He comes up with a theory that exhibits certain essential features. Then he tests that theory against experimental results. If it had been found wanting, it would be discarded, but in practice it worked every time.

The character Sagredo is a little doubtful at first of Salviati's assertion that under the effect of gravity, heavy objects start off moving very slowly and gradually accelerate. Here we see a problem because of the lack of suitable instruments to measure what is happening. If, say, you let go of a hammer, it seems to be moving at quite a speed as soon as you can register that it is moving. But Salviati reassures Sagredo with a thought experiment.

He imagines dropping a heavy object onto a stake that is pushed into the ground. Drop the object from 4 cubits (around 1.8 meters or 6 feet) and it will drive the stake in a considerable distance. But drop it from very close to the top of the stake (the thickness of a leaf) and the result will be imperceptible. The heavy object weighs the same in both cases, but when it is dropped a very short distance it has very limited impetus because it is moving so slowly.

For Galileo, there is none of the classical concept of something craving to reach the center of the Earth because it is its natural tendency. With a physicist's eye, Galileo sees things in terms of forces. He doesn't get everything right. Like all his contemporaries he thinks that something thrown upward, say, has an "impetus," an upward force on it that gradually declines as it moves, where actually there is no upward force as soon as the object leaves the hand. But he is clear that acting downward is the "force of gravitation."

What Galileo is also clear about, though, is that he has no intention of trying to explain why or how gravity applies a force and causes bodies to accelerate. Some people, he tells us, explain it by attraction to the center, others by repulsion between the very small parts of the body and others by some kind of stress in the surrounding medium. There isn't a clearly understood mechanism, as far as Galileo is concerned, of how gravity applies its force, but his experimental observations make it clear that it does.

Despite that lasting image of balls dropping from towers, most of Galileo's experiments with gravity were done using inclined planes—sloping pieces of wood. By rolling balls down the planes under the force of gravity he was able to study its effects in a more controlled and measurable way. He describes cutting a channel in a piece of wood, making it "very straight, smooth, and polished, and having lined it with parchment, also as smooth and polished as possible, we rolled along it a hard, smooth and very round bronze ball." He was doing his best to minimize friction.

He initially toys with the reader by having one of his characters suggest that speed increases in proportion to the space traversed—so, for example, "the speed acquired by a body in falling four cubits would be double that acquired by falling two cubits."[14]

But this is an error, Galileo points out, because the increase in speed is dependent on the passage of time, not the distance covered, and because as the body gets faster and faster, the distance it covers in that time will be bigger and bigger. He then goes on to show that distance traveled when accelerating under gravity (or any uniform acceleration) depends on the square of the time taken, but is not dependent on the weight of the body.

Making such measurements with falling bodies would be difficult using the technology available to Galileo. His quick-and-dirty timing mechanism was to count heartbeats, while in his more detailed experiments he used a water clock which let a fine jet of water out of a large vessel, and weighed the water released on different runs to compare timing.

Sometimes measurements were taken by ear rather than eye. Perhaps influenced by his father the musician, in some experiments Galileo placed wires across the sloping piece of wood, which could be heard when the ball passed over them. He then spaced out the wires until the sounds of the ball passing over them were evenly spaced. The result gave a simple measurement of how far the ball had passed in a unit of time.

Galileo had taken huge strides in laying the groundwork for a better understanding of gravity. His observations of Jupiter's moons and other discoveries with his telescopes had destroyed the main arguments for keeping an Earth-centered universe with its associated theory of gravitation. He had disproved Aristotle's contention that heavier objects fell faster, and with detailed experiments had shown the nature of the acceleration due to gravity. It was time for the mantle of master of gravitation to be passed on.

If you were to ask a physicist the one name they most associate with gravity it would be Einstein (of whom much more later), but for everyone else the key name is Isaac Newton. He shares with Einstein joint top position as most famous scientist in history, and over the years his life story has had the kind of massaging that is usually reserved for Hollywood stars.[15]

Isaac Newton was born in his family home in the village of Woolsthorpe, in Lincolnshire in the North East of England. The house (still there to visit) was called Woolsthorpe Manor, which makes it sound like a grand country mansion, but in truth it was little more than a solid stone-built farmhouse, well suited to a genteel farming family.

If it hadn't been for his mother's family, Newton would probably have stayed on the farm and never done the great things he did. But when Hannah Ayscough married Newton's father (also Isaac), she brought a higher social standing with the expectation of a university education for any children. We don't know what Isaac senior thought on the matter, though. Newton's parents were married in April 1642 and less than a year later Newton's father was dead.

It is entirely possible that Newton was conceived before the wedding as he was born on Christmas morning 1642. We don't know for certain, because Newton certainly engaged in some rewriting of history (for example giving an earlier date for his parents' marriage when he was knighted). It is often pointed out that he was born the same year as Galileo died, though this coincidence depends on using the dating of the time. England had yet to switch to the modern Gregorian calendar—by modern dating, Newton was born in January 1643.

After all too few settled years, Newton's life was disrupted in 1646 by his mother's marriage to Barnabas Smith, rector of the nearby village of North Witham. Hannah moved to live with her new husband, but Newton was left behind at Woolsthorpe with his grandparents. It is probably unwise to attempt to psychoanalyze Newton at this distance, but it doesn't seem unreasonable that Newton's striking independence—he would be a loner all his life—was influenced by this early desertion.

His mother did return in 1653, when her second husband died, but she brought with her three children from her new marriage. Now Newton had a very different disruption—to the peace of his home—and it seems highly likely that both he and his mother found it something of a relief when the 11-year-old boy was sent off in 1654 to stay with a family called Clark in the nearby town of Grantham, so Newton could attend the King's School there.

Mr. Clark was an apothecary—a role that fit somewhere between a modern day pharmacist and a clinician. His shop was a home to a wide variety of chemicals and was a place where Newton was likely to be exposed to the scientific ideas of the day. At the same time, Newton was beginning to do well in school, and by the time he was sixteen it was obvious that he was destined for a university career. At this point his mother decided he should come back to the farm. She wanted him to run the family business, rather than concern himself with too much education.

Newton did not go quietly. When he was given jobs around the farm he would get the servant to do them and return to his studying. When he was sent to market in Grantham he would sneak back to the Clarks' to pick up news of the latest scientific developments. His headmaster at the Grantham school, Henry Stokes, heard just how important learning was to young Isaac. With the support of Newton's grandfather (himself a Cambridge graduate), Stokes managed to change Hannah's mind. In June 1661 Newton made the journey to Trinity College, Cambridge.

His mother had no intention of making it easy for him. A student with his background would be expected to be a pensioner, a student position that was funded to have a reasonably comfortable life, but instead his mother insisted he took the role of a sizar, where in place of fees he was required to act as a servant to other students. It didn't stop Newton, though. The move to Cambridge freed up an already expanding talent and he would never look back.

Even the advent of plague, forcing him to abandon Cambridge for 2 years, would not stop his rise. In fact it seems to have helped. Along with London, Cambridge had a serious outbreak of the bubonic plague in 1664 and the university was shut down. Most of his contemporaries used the 2-year break for entertainment, but the 23-year-old Newton, ever serious, instead took the opportunity to think, laying the groundwork for all his great future work, from his explanation of light to his work on gravity.

Think of Isaac Newton and gravity and one thing springs to mind: an apple, falling from a tree, quite possibly hitting Newton on the head. This near-mythical image inspired the first logo of Apple Computer, for example, which looks like a woodcut of Newton sitting under a tree with a glowing apple above him, ready to fall. (The name of the company was apparently suggested by Steve Jobs because it was his favorite fruit and because he had worked in an orchard, but presumably the idea of Newton's breakthrough moment came up at some point before the logo was created.)[16]

There has been so much effort of late to dismiss the apple-on-the-head story, that there's a real danger of throwing the baby out with the bathwater. There is no suggestion whatsoever that Newton was inspired to liken the gravity that pulls down apples with the force that keeps planets in place when he was brained by a Red Delicious. However, it does seem that an apple was involved in the initial inspiration, as we have the word of Newton himself (via a third party).

The apple incident is recorded in a book called *Memoirs of Sir Isaac Newton's Life,* written by the antiquarian William Stukeley, younger than Newton, but still able to interview Newton before the end of his life. We can get a particularly good flavor of this meeting, as the manuscript of Stukeley's book has been put online by the Royal Society.

Stukeley "paid a visit to Sir Isaac" on April 15, 1726, at his lodgings in Orbol's Buildings, Kensington (London), where he dined with Newton and "spent the whole day with him alone." Then comes the key passage:[17]

> After dinner, the weather being warm, we went into the garden, and drank thea under the shade of some apple trees; only he and myself. Amidst other discourse, he told me, he was just in the same situation, as when formerly, the notion of gravitation came into his mind. Why should that apple always descend perpendicularly to the ground, thought he to himself; occasion'd by the fall of an apple, as he sat in a contemplative mood.

Of itself this is interesting, and is enough to give rise to the legend, but it seems quite a leap to get to the idea of gravity keeping the orbiting planets in their places. Yet read more of Stukeley's account and you get a much clearer picture—admittedly in no sense technical as Stukeley was writing for a popular audience—but describing the chain of thought that led from apple to universal gravitation. Of the apple, Stukeley observes:

> Why should it not go sideways, or upwards? But constantly to the earths center? Assuredly the reason is, that the earth draws it. There must be a drawing power in matter. The sum of the drawing power in the matter of the earth must be in the earths center, not in any side of the earth. Therefore does this apple fall perpendicularly, or towards the center. If matter thus draws matter; it must be in proportion of its quantity. Therefore the apple draws the earth, as well as the earth draws the apple.

This seems very straightforward to our contemporary view of gravity, but there would have been a number of surprises for any of Stukeley's readers who weren't familiar with Newton's work, particularly the concept that the apple pulls the Earth, not just the Earth pulling the apple. The next part shifts the viewpoint to planetary motion:

> That there is a power like that we here call gravity which extends itself through the universe and thus by degrees, he

> began to apply this property of gravitation to the motion of the earth, and of the heavenly bodys: to consider their distances, their magnitudes, their periodical revolutions: to find out, that this property, conjointly with a progressive motion impressed on them in the beginning, perfectly solv'd their circular courses; kept the planets from falling upon one another, or dropping all together into one center.

To reinforce just how significant this all was, Stukeley makes it clear that Newton's work is the key to understanding the universe, and, with a patriotic flourish, puts British thinking ahead of its continental neighbors:

> And thus he unfolded the universe. This was the birth of those amazing discoverys, whereby he built philosophy on a solid foundation, to the astonishment of all Europe.

Stukeley doesn't tell us where the apple tree was that inspired this thinking, but it is generally assumed to be a tree at Newton's old family home, Woolsthorpe Manor, where there is a 400-year-old apple tree of the Flower of Kent variety that dates back to Newton's time, placed squarely in view of his bedroom window.

This tree has become the modern equivalent of a medieval shrine. When people pay a visit to Woolsthorpe, they inevitably head for the tree and want to touch it or be photographed with it, just as pilgrims might touch the shrine of a saint. The tree survived being partly blown down by a storm in 1820, but the Newtonian pilgrims have put it in danger, trampling the soil around it so much that water was not getting through to the roots. It has now been surrounded by a willow barrier to keep overenthusiastic tourists at a respectful distance.[18]

Newton, then, seems to have worked out some of his ideas on gravity during those remarkable 2 years in Lincolnshire when the university at Cambridge was closed, but he was often late to publish his ideas, perhaps distracted by other interests along the way, and it was not until 1687 that Newton's masterwork that established his law of universal gravitation was published under the auspices of Astronomer Royal and Newton supporter Edmond Halley.

Halley also penned an introductory ode that is published in the book, not something you would expect these days in a science textbook. It is turgid stuff but its opening lines make it clear the significance of what was to come:

> Behold the pattern of the heavens, and the balances of the divine structure;
> Behold Jove's calculations and the laws
> That the creator of all things, while he was setting the beginnings of the world, would not violate;

These lines are particularly interesting when you consider that Halley was unusual for his time in being a vocal atheist—he might not be denying the existence of God, but he is making it clear that, as far as he was concerned, Newton's work was describing natural laws that even the creator had to follow. He finishes with a paean to Newton himself:

> O you who rejoice in feeding on the nectar of the gods in heaven,
> Join me in singing the praises of NEWTON, who reveals all this,
> Who opens the treasure chest of hidden truth,

The book was originally written in Latin, titled *Philosophiae Naturalis Principia Mathematica (Mathematical Principles of Natural Philosophy),* though it is usually referred to for convenience as the *Principia.*[19] This is far more than a treatise on gravity. It encompasses all Newton's idea on mechanics, forces, and moving bodies, including his famous three laws of motion. But unlike other works on these subjects, Isaac Newton's greatness was to see further. To take these laws of motion that applied to everyday objects on the Earth and to apply them to the movement of the hitherto unfathomable bodies that inhabited the universe.

Halley had encouraged Newton to write something explaining how orbiting bodies might move, and from earlier discarded versions of the manuscript, this seems to be all Newton originally intended, but his scope grew until it reached the final magnificent (if sometimes impenetrable) form.

Newton was not working in isolation. Kepler's laws of planetary motion were already more than 60 years old when the *Principia* were published. These accurately described the movement of the planets, but did not explain why they did not fly off into the void. The first law covers the shape of planetary orbits. It says that the orbits trace out the squashed circle that is an ellipse, with the Sun at one of its foci. (The two foci of an ellipse are the equivalent of the center of a circle. If you take any point on the ellipse and add up the distances from that point to the two foci, it will be a constant.)

The second law indirectly covers the speed of a planet as it travels around its elliptical orbit. It says that a line from the planet to the Sun will produce shapes of equal areas in a fixed amount of time. What this implies is that when the planet is closer to the Sun, making the area swept out smaller for any particular angle, it must be moving faster.

The most complex sounding of the laws is the third, which Kepler developed quite a while after the other two. It says that the square of a planet's orbital period is proportional to the cube of its semi-major axis. So if you square the time it takes a planet to go around its orbit, it will always have the same proportion to the cube of the distance that is half the length of the diameter passing through both foci of the ellipse.

Kepler's work was one part of the foundation on which Newton built his theory. A second contribution probably came from one of Newton's great rivals, the Dutch scientist Christiaan Huygens. Newton was anything but happy with Huygens's theory of light, published in 1878, which argued that light was a wave, where Newton was convinced that light traveled as particles he called corpuscles. But Newton does seem to have been positively influenced by Huygens's earlier *Horologium Oscillatorium,* which shows that the force on an object traveling steadily in a circle is proportional to its velocity squared divided by the radius of the circle.[20]

Combining this value with Kepler's third law soon made it clear that the force acting on bodies to keep them in orbit must be an inverse square law—a force that depends on the square of the distance from the Sun in the case of the planets—and a force that gets rapidly smaller as the distance increases. Here was another piece to Newton's jigsaw, while a further component came from his arch rival at the Royal Society, Robert Hooke.

Hooke was curator of experiments, and later secretary of the Royal Society. A few years older than Newton, he had a much broader range of interests and a social ease that Newton could never match. Where Newton spent much of his time alone in his rooms, Hooke was part of the coffee house set and a reputed womanizer. Hooke thought Newton self-centered and inept; Newton considered Hooke a shallow dandy.

From their first contact, Hooke and Newton rubbed each other the wrong way. Newton had submitted a paper on light to the Royal Society and Hooke gave it a poor review, in part because he spent little time looking at it. Yet Hooke was a superb scientist in his own right, and he came up with the counterintuitive idea that an object in orbit takes the path it does because of a combination of two different movements. The object is both falling freely toward the body around which it orbits, and moving in a straight line at a tangent to the orbit it describes.

This can be seen with dramatic consequences in an orbiting space station. The crew members float around weightless. But this weightlessness is not because there is no gravitational pull at the distance the station orbits. After all, if there were no pull, the space station would not stay in orbit. The crew is weightless for the same reason that people in a dropping elevator or a "vomit comet" aircraft are weightless. Such aircraft climb to a considerable height, then drop in free fall for about 20 seconds at a time before repeating their parabolic flight, allowing the passengers to experience weightlessness. Similarly, the crew on a space station are in free fall toward the surface of the Earth. But the clever thing about being in orbit is that, though they are always falling, the sideways motion is just the right amount to make sure that they always miss.

We know that Newton got this idea from Hooke because he told Hooke that he had never heard of this "hypothesis" before.[21] Newton also saw Hooke's suggestion that there was an inverse square law relationship between the Sun and the planets. When the *Principia* was published, Hooke claimed that Newton had stolen his ideas, yet in the letters between the two, it seems rather that Hooke had gone as far as he could and was encouraging Newton to take his ideas further—as Newton certainly would.

Before we see Newton's attack on gravity in full flower, it is useful to take a dip into his definitions. In a short section of the book he describes eight terms that are "less familiar" than words like "time, space, place, and motion," which he believes are familiar to everyone. Some of the words he defines are rather obscure variants on force that we don't give the same separate importance simply by moving the body from location to location. Newton was looking instead for an inherent measure of the amount of matter in an object, a measure that would be unchanged wherever that object was. It was this that would become known as mass.

We confuse weight and mass because we have adopted the same units and values for the two quantities. Both are measured in kilograms (or pounds) and on the surface of the Earth, the mass and weight of an object have the same value. This is an entirely arbitrary decision, which makes calculations easier, but produces confusion as soon as you cease to think parochially.

Newton decided that his "quantity of matter" (which he eventually reduced to "mass," a term I will use from now on, because it is less clumsy) was a result of a combination of the density of an object—how tightly stuff was crammed into it—and its volume. Mass was, he believed, proportional to weight in any specific circumstance, but unlike weight, the mass of an object would not vary from location to location, on or off the planet. Similarly it would not change when you heated the substance or otherwise manipulated it physically, provided you did not rip bits off it.

It might seem that Newton was employing a circular argument in defining mass, because he thought of it as a function of density, and we now define density as mass per unit volume. However, he was using density in a more direct fashion, describing how much of the particular stuff being studied there is in a certain volume. As he was a believer in atoms (even though their existence in the modern sense was unknown at the time) he may well have had in mind the density as simply the number of atoms

After its introduction, the third book proceeds through Newton's analysis of planetary motion and of that other striking real-world result of gravitation, the tides. The book finishes with a "general scholium," a section that pulls together the contents of the work. It is here that Newton presents us with one of the most famous phrases in the *Principia*, one that will be crucial in the consideration of just what gravity *is*, something we will return to in the next chapter.

To modern eyes, Newton's book is very wordy, and relies to a surprising degree on geometry when dealing with aspects of gravity. However at the heart of the book is a formula that, though never explicitly used, can be deduced from Newton's various calculations and diagrams. It is now known as Newton's law of universal gravitation, and is usually presented in the form

$$F = Gm_1m_2/r^2$$

Here we see the force of gravity (F) acting between two objects laid bare. It is proportional to the inverse of the square of the distance between the objects (r)—as the distance gets bigger, the force gets weaker, and this happens quicker and quicker as you get further away. It depends on the mass of each of the two objects (m_1 and m_2) and to complete the formula we have "G" which is a constant value that allows us to work out any one of the other items given the rest.

There was a catch, though. The powerful revelation of Newton's study of gravity was that it was an inverse square law—the force got bigger as the square of the distance between the objects gets smaller. But how did you measure the distance an object is from a large body like the Earth? Clearly the "r" in the equation couldn't be the distance from the surface of the planet or the force of gravity there, felt by a human being standing, for example, would be divided by 0—it would be infinite.

Imagine dividing the Earth up into lots of person-sized chunks. Each of these chunks would have a tiny attractive force on the Moon, pulling toward that specific chunk. But overall, Newton had to be sure that the result would be the one we tend to assume—that the attraction was toward the center of the Earth. He managed to prove this would be the case for spherical bodies—which takes in the planets and the stars, though there are some smaller moons and asteroids which are anything but spherical, and for them locating the center of gravity becomes a little more tricky.

Newton didn't have a value for G and only ever presented his law of gravitation by comparing two forces and how they varied. In fact it was not until 1894 that the British physicist Charles Vernon Boys came up with the concept of the universal gravitational constant, G, to provide the final version of the formula.[24]

You may remember a formula for the force of gravity from school that was much simpler

$$F = mg$$

This is based on the assumption that you are at a fairly constant distance from the center of the Earth. In that case, you can combine G, m_1 (the mass of the Earth), and $1/r^2$ into the single constant g, which represents the acceleration due to gravity—around 9.8 meters per second per second (roughly 32 feet per second per second). It's just a simplified version of the full equation.

The first attempts to measure the fundamental attraction between masses would be done using a pendulum, just as Galileo had done before. But this was no ordinary pendulum. The experiment was undertaken by Henry Cavendish, the eighteenth-century British physicist. Cavendish was the son of a wealthy aristocratic family and had the funds to support a highly solitary life dedicated to science. He worked on gasses and electricity but also made an important step forward in measuring the force of gravitational attraction.

Working in the 1790s, Cavendish built an apparatus featuring a torsion pendulum. In this kind of pendulum, instead of swinging in a straight line, the pendulum twists back and forward. Think of what happens when you twist a child's swing around on its ropes. It rotates one way, past the "untangled" point, then rotates back the other way. A torsion pendulum is like this, rotating back and forth because of the force in the twisted cable that supports it.[25]

On the end of Cavendish's pendulum was a bar with a ball at each end. There were other larger, heavy balls placed near the swinging bar, and the instrument was used to measure how the gravitational attraction between the large balls and the pendulum bar influenced the back-and-forth twist of the pendulum. Considering the crudeness of his technology and the tiny force involved, it is remarkable that Cavendish achieved any results at all.

Cavendish used this technique to estimate the density of the Earth, but very sophisticated modern versions of the experiment are used to determine the universal gravitational constant G to greater accuracy than was possible in the eighteenth century. In 2000, for instance, a torsion pendulum was built at the University of Washington by Jens Gundlach and Stephen Merkowitz. Here a piece of Pyrex, acting as the bob of the pendulum, was subject to attraction from very accurately made 8 kilogram (17.6 pound) steel balls. Their measured value of G was $(6.674215 \pm 0.000092) \times 10^{-11} m^3 kg^{-1} s^{-2}$.

This sounds impressive, but it is not anywhere near as accurate as our knowledge of the other fundamental constants of science like the speed of light or Planck's constant. Making such a measurement is hugely tricky, in part because gravity is such a weak force compared to electromagnetism, so it is very easy to be misled by stray influences. Measuring G these days is not something you do over a few days—a typical experiment will take several years to complete.

Although Newton never came up with the formula involving G, his work on gravitation was to have a huge impact—yet there was an issue that even he would acknowledge. He had no clue as to how gravity did its stuff. And any attempt to explain it that fit with the picture he had developed in the *Principia* seemed to require a particular dubious prescientific concept. The mystical "action at a distance."

CHAPTER FIVE

ACTION AT A DISTANCE AND OTHER GRAVITATIONAL MYSTERIES

> *But shall gravity be therefore called an occult cause, and thrown out of philosophy, because the cause of gravity is occult and not yet discovered? Those who affirm this should be careful not to fall into an absurdity . . .*
>
> —*Mathematical Principles* (1729)
>
> Roger Cotes

Action at a distance is a simple concept, but one that doesn't fit into the Enlightenment picture of a universe where everything interacts mechanically, like the components of a vast cosmic clock. It is more a thing of magic and mystery. Action at a distance means that I click my fingers here and something happens elsewhere. Somehow, mystically, the *need* for action to occur in one place is translated into actual action elsewhere without anything traveling through the intervening space.

If you ignore forces like gravity, this idea of action at a distance runs counter to both everyday experience and common sense. Generally speaking, if we want to make something happen remotely, we need to send something physical across the space in between to make it happen. If you see a can on a distant wall and want to knock it off, you have to throw a rock across the space between you and it. It isn't enough to simply will it to fall off.

Although there is nothing obvious moving, this is also the case when I make a sound and that sound reaches someone else's ear. To start the sound off, I create a pressure wave in the air with my vocal cords. This crosses the intervening distance, and eventually comes into contact with the other person's eardrum. The wave, formed from rhythmical movements in the air molecules, crosses the room from A to B, making the connection between source and target. Similarly, if I control something remotely by electricity, an electromagnetic wave and a stream of electrons make the journey across the gap. But gravity seems to act at a distance with nothing intervening.

This kind of remote influence is something that human beings inherently react against. When we see a magic trick where the magician appears to move something remotely, we assume there are wires or some other invisible but solid connection between the two. Even babies have been shown to be suspicious of action at a distance, paying more attention when it appears to happen than they would normally. Yet with Newton's introduction of gravitational attraction, he seemed to invoke action at a distance.

In the final "general scholium" of *Principia,* Newton tells us that the motion of the heavenly bodies (and the tides) are down to gravity, but that he has not decided on a cause for this gravitational pull. Because it acts in proportion to the total quantity of matter, rather than the size of an object's surface, he deduces that this "gravity" thing can't be mechanical in the conventional sense. However, he does not have a solution and—in the original Latin of *Principia*—he comments "Hypotheses non fingo," which is traditionally translated as "I frame no hypotheses."

Newton wasn't saying that he never uses hypotheses—in fact, as we have seen, book 3 of *Principia* starts with a bunch of hypotheses—but rather that he is not prepared to make up some hypothesis for the cause of gravity. That word "fingo" is important as it doesn't mean something neutral like "propose." The Latin word (like the English "frame") is derogatory, suggesting making something up, inventing fictions, rather than conceiving in a thoughtful, scientific fashion. This is "frame" in the sense of being framed for a crime.[1]

Newton concludes by saying that it is enough that gravity really exists and acts according to the laws he has described. In effect, as long as the math works to predict what happens in the world it doesn't really matter whether or not we understand why that outcome occurs. This is a surprisingly modern approach—when twentieth-century physicists developing quantum theory took a similar view, they were sometimes criticized for not having a good enough reason for *why* the math worked.

Some of Newton's contemporary critics were not impressed by his apparent distancing himself from the causes of the attractive forces of gravity. For them it was enough that he said that there was an attraction at a distance between bodies to accuse him of stepping beyond science into mysticism and of dealing with occult forces.

This concern about Newton's theories wasn't helped by that word he used, "attraction." We are now so familiar with the idea that the Earth attracts the Moon and vice versa, or that a magnet attracts a bit of metal, that we don't really think about the word we are using. It is perhaps clearer if we say something is "attractive." The word was not originally about a physical force, but about an animal reaction. Newton's contemporaries were concerned that he seemed to be basing his gravitational theory on the fact that the Earth and the Moon fancied each other, that the Earth attracted the Moon the way an attractive person picked up followers.

The Dutch scientist Christiaan Huygens, always something of a terrier of science, worrying away at Newton's work, was not impressed by this use of attraction. He dismissed the "theories [Newton] builds upon his Principle of Attraction, which to me seems to be absurd."[2] Similarly, Newton's other great rival, the mathematician Gottfried Wilhelm Leibniz, attacked the idea that two bodies could be attracted toward each other, calling it a "return to occult quantities and, even worse, to inexplicable ones."[3]

The feeling was that Newton had taken a step backward. It might seem obvious to us today that gravity is a force that works at a distance, but for his contemporaries this harked back to a pre-Enlightenment era when "occult" (which is to say, hidden) forces were assumed to be at work and angels were considered an acceptable cause for natural phenomena. The whole thrust of scientific thinking in Newton's era, typified by Descartes' work, had been that everything was now explicable through mechanical means. They had put away childish occult fantasies.

According to such mechanical theories, if something moved, then it was a result of pressure, and if that pressure acted at a distance it was as a result of an invisible chain of objects each pressing on the other. There had to be a physical link. For Descartes, the planets were kept moving in their orbits by a series of vortices in an invisible fluid. Huygens had taken this approach and given it a better mathematical basis. But Newton seemed intent on throwing away the framework, leaving an unsupported force at work.

Newton denied that his approach did away with a mechanical basis for gravitation. While the *Principia* carefully stayed away from an explanation, he believed that his gravitational work was entirely compatible with a fluid-based explanation of gravity. He made his feelings clear in a letter to classical scholar and theologian Richard Bentley. Written soon after the *Principia* was published, the letter declared that no one in their right mind could believe it was possible for a body to "act where it was not."[4] His use of the term attraction was a way of describing the result, not a suggestion for the cause.

Yet if Newton believed this, how did he conceive that gravity enabled the Sun to influence distant planets and comets? After toying with concepts like electricity, he eventually settled on a similar approach to Huygens, that massive bodies influence an "aetherial medium," an invisible material that filled space and that provided the indirect contact needed for one body to influence another, just as air, in a different way, enabled the sound produced at one place to influence an ear elsewhere. Despite the attacks on his concept of attraction, Newton was equally uncomfortable with action at a distance.

Although a good number of contemporaries sniped at the no-strings-attached attraction of Newton's gravitation, they could not deny the elegance of the mathematics or the effectiveness of the solutions in matching reality. While Newton's approach gave no explanation, it certainly delivered the goods. But it didn't stop further theories being constructed to give a good, sound mechanical explanation for the behavior that Newton's law of gravitation predicted.[5]

Like Newton's own, these theories relied on some sort of medium existing throughout space, whether that same "ether" that was thought to carry the waves of light or some other substance. The longest lasting of the theories was originally developed by the Swiss mathematician Nicolas Fatio de Duillier, but later (and possibly independently) conceived by his countryman, physicist Georges-Louis Le Sage. Variants on this theory would be produced all the way up to the beginning of the twentieth century, most notably by William Thomson (Lord Kelvin), though Le Sage published his ideas in 1748.[6]

Most ether-based ideas considered the aether to be a continuous fluid of some sort, but the de Duillier theory gave the ether a Newtonian twist. Newton believed that light was made up of particles, or corpuscles as he called them. In the mechanical theory of gravitation that de Duillier and Le Sage proposed, the ether itself had a corpuscular structure. These particles weren't static, but instead flew around in all directions, smashing into solid objects. Just as we now know that air molecules slamming into objects causes air pressure, it was thought that these gravitational corpuscles would influence the bodies with which they collided.

If a single object was isolated in space, it would not move as a result of the pressure from the corpuscles because they came at it from all directions in equal quantities. But think what would happen if two bodies occupied space relatively near each other. Each body would be shaded from corpuscles traveling from the direction of the other, just as the Moon shades the Sun and causes an eclipse. There would be fewer corpuscles hitting the bodies on the sides facing the other. The result? The two bodies would feel a net pressure toward each other. They would be attracted.

What is particularly neat about this model is that it automatically generates the inverse square law that is so central to Newton's workings. But over the years it was proposed, many were doubtful about it, because the solution seemed as bad as the problem. This was a mechanical approach that did away with action at a distance, but it required the universe to be full of streams of invisible undetectable corpuscles that had a combined effect that was powerful enough to keep the planets in their courses.

The corpuscular theory required a major imaginative leap. Where did these particles come from? What drove them with such power? The theory produced as many questions as it did answers. Yet flawed though it was, it was probably the closest there was to an explanation for the action of gravity before Einstein came along. What's more it is one that would probably feel more comfortable to modern physicists, used to vast quantities of particles flowing through space, than it ever could to Newton's followers.

It is in part because of its nature as an apparent action at a distance that gravity so often fools us by allowing the unexpected to happen. A simple example would be something that Galileo knew, but still causes confusion today. Imagine that I have two identical bullets, one in my hand, the other loaded into a gun. I fire the gun horizontally, avoiding any potential targets, and simultaneously drop the second bullet from the same height. Which bullet will drop to the Earth first?

Our natural tendency is to assume that the falling bullet will be first to reach the ground. There seem to be two reasons for this. We know roughly how quickly things fall. A bullet, dropped from shoulder height, would hit the ground in around a second. It seems very unlikely that a bullet fired from a gun would reach its maximum distance so quickly—but if it is fired on the level, that is exactly what will happen.

The other reason we fight against the notion that both bullets fall at the same speed is a mistaken feel for a property of matter that is a kind of super-inertia. It seems that when something shoots forward at high speed it shouldn't fall. Think of a cartoon characters running off the edge of a cliff, and only beginning to fall once they realize there is nothing beneath them. It is almost as if we expect the bullet to have some similar ability to stay at the same altitude for some time before it begins to fall because of its forward motion.

It is just possible that a bullet could be given some lift by the air, keeping it up a little longer than expected, but the shape of a bullet and its density make it unlikely that it will gain much benefit. The chances are that the match in time to fall will be good. If you ignore the air, the only force acting up or down on the bullet is the force due to gravity. And that is exactly the same for both bullets. Each will accelerate downward at around 9.8 meters (32 feet) per second every second.

While we are on the subject of bullets and gravity, we've all seen people on TV celebrating by firing guns into the air. With a little consideration of how gravity acts, this is a truly terrifying act. Consider the forces acting on the bullet. As the gun is fired, the bullet receives a massive push upward, accelerating it to perhaps 600 miles per hour (268 meters per second). But as soon as the bullet leaves the gun, excepting relatively light air resistance, the only force acting on it is gravity.

As the bullet heads upward it feels the pull of gravity toward the ground. Although still traveling up it is accelerating downward, so the result is that it slows. In a brief while it will come to a stop. It might seem, as the bullet is poised at the top of the trajectory, it has balanced forces acting on it—but no, gravity is still there, still pulling downward, providing that relentless acceleration. Soon the bullet is plummeting.

Here is where it gets interesting. There will be some slowing from air resistance, but the terminal velocity of a bullet is still as much as 150 meters (490 feet) per second. By the time the bullet returns to ground level it is traveling very quickly. What we have here is a lethal bullet, traveling at several hundred miles per hour. Heading straight down toward the heads of the people who shot it upward in celebration. A hail of bullets will head back to them out of the sky.

Of course, many bullets will be shot at an angle and fall to the ground harmlessly, and it is possible that a bullet will begin to tumble and lose velocity from turbulent flight, but there is reasonable evidence both from tests and from a good number of news reports of deaths that this practice is potentially deadly.[7]

Another gravitational source of confusion arises from its effect on the tides. Newton was, as we have seen, the first to come up with a sensible link between the tides and gravity. Galileo had also made an attempt to explain the tides, but in this he made a huge mistake. He tried to use the tides as a significant, but woefully incorrect argument in favor of the Earth going around the Sun in his life-threatening book *Dialogue Concerning the Two Chief World Systems.*[8]

Galileo suggested that the tide was a side effect of the Earth's passage around the Sun. As the Earth rushed around the Sun, he thought it only reasonable that the water behind the Earth's motion would bulge out. This idea was so central to Galileo's argument that his working title for his book was *Dialogue on the Tides*, and he originally submitted it under the title *Dialogue on the Ebb and Flow of the Sea.* Unfortunately this model would only produce one tide a day, but Galileo attempted to argue this problem away by introducing interaction between the rotation of the Earth and its motion around the Sun to produce the second tide.

By the time Galileo produced his theory there were already plenty of suggestions of the Moon's involvement in the tide. Those who kept detailed records of the tides could hardly fail to notice that they shifted with the Moon's motion around the Earth. Galileo mocks this, saying that a "certain prelate" has published a tract saying that the Moon "attracts and draws up towards itself a heap of water which goes along following it . . ." Others, he tells us say the Moon can "rarefy the water by its temperate heat."

This "rarefaction" by the Moon reflects the medieval understanding. Roger Bacon, writing in the thirteenth century, tells us that Arab scholars are of the opinion that the Moon is the cause of the tides. Bacon describes rays from the Moon causing vapors to rise from the depths of the sea "like swelling bottles, and overflowing waters of the sea, so that they are driven from their channels."[9] The idea of linking the tides to the Moon went back to ancient times, but philosophers had struggled to come up with a legitimate reason for the Moon to have such a powerful influence.

Newton's newly discovered universal gravitation seemed an ideal explanation for the tides. If the Sun's gravitational pull was enough to keep the Earth on its path, then surely it would have some influence over the water on the surface of the Earth. Similarly, the Earth and the Moon were bound together by a gravitational pull, and the Moon's gravity should influence the sea.

Newton spends quite a considerable time explaining how the Moon and Sun produce the various tidal movements. He stresses that it is a combined effect—although we might simplify the picture by saying that the Moon's influence produces the two daily tides, while the Sun produces the annual variations from spring tide to neap tide, in reality it is always a combination of the two that produce the familiar pattern.[10]

It might not immediately seem obvious that the Moon should produce two tides (if we temporarily eliminate the Sun to keep things simple). Think of the position of the Moon relative to the Earth. The seas on the same side of the Moon as the Earth will be attracted toward the Moon, so the tide will be high on that side. By comparison, on the far side of the Earth, the attraction of the Moon will be weakest, so the water should bulge away from the Moon causing a second tide. However this tide would be very low indeed if this were the only factor—much lower than it really is.[11]

Newton had identified one key element to the tides, but he didn't get the cause totally right. If it were left to the attraction of the Moon and Sun alone there simply would not be enough gravitational attraction to cause the tides we observe, nor would you get the more complex variations in tidal flow. This is because the Earth's rotation and particularly the rotation of the Earth/Moon system also have an influence.

The Earth's rotation is fairly obvious, but the Earth/Moon

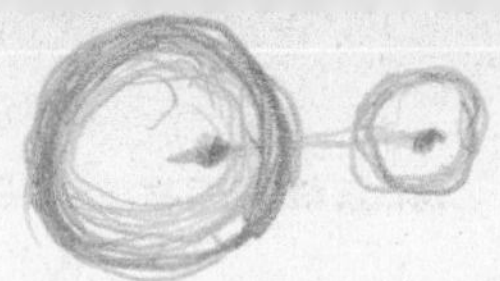

system needs a little explaining. We tend to think of the Moon rotating around the Earth, but that is a one-sided picture. Instead, both the Earth and Moon rotate around their joint center of gravity, a point that is within the Earth (because the Earth is far more massive than the Moon), but around 3/4ths of the way to the Earth's surface. This means that the Earth undergoes a gyration like a fairground ride that adds to the amount of force applied to the oceans to power the tides. This is the main reason for the tidal bulge on the opposite side of the Earth to the Moon.

There are tides in the planet itself as well as the oceans. We tend to think of the Earth as being a solid lump of rock, but it is mostly molten, and even the solid outer layers can shift under the influence of gravity. The tidal movement in the Earth can be as much as 20 percent as large as the movement in the sea (shifting about, 30 centimeters [12 inches] every 12 hours) but we don't notice it because there is nothing with which to make a comparison.

If there were water on the Moon, we would see tides in a whole new light, because the tidal forces that the Earth produces on its satellite are dramatic indeed. In total, the Earth provides tidal forces 80 times the size of those felt as a result of the Moon's gravitational pull. This may seem too large if you know that the gravitational pull on the surface of the Moon, as demonstrated by moonwalking astronauts, is around 1/6th that of the surface of Earth.

The reason the surface gravity on Earth is 6 times bigger than the Moon's, but the tidal forces it produces are 80 times bigger reflect the fact that the force of gravity depends on the distance from the center of mass. The Earth's radius is 3.6 times bigger than that of the Moon, so someone on the Moon is 3.6 times closer to the Moon's center. This means they will feel around 13 (3.6 × 3.6) times as much force for the same amount of mass. So though the Earth is 80 times more massive than the Moon (and generates 80 times the tidal force), the surface gravity is only 80/13 times bigger.

Although we can't see water sloshing around on the Moon to demonstrate how strong the tidal forces are, just looking at it over time provides clear evidence of the Earth's tidal forces at work. Whenever we look at the Moon, we see the same face. It changes shape, waxing and waning as the angle of light hitting it from Sun changes, but the face of the Moon that points toward us is always the same one. We never see the far side (often called the dark side, though it is no darker than the side we see).[12]

At first sight this is merely an amazing coincidence. And astronomical coincidences do happen. The Moon itself features in one when it appears to be almost exactly the same size as the Sun. Despite being around 400 times smaller, the Moon is also around 400 times closer, so it covers the Sun almost exactly in a total eclipse. There is no reason for this, it is just chance. However the "coincidence" of the same side of the Moon always facing us is no coincidence at all.

For this to happen, the Moon has to spin on its axis at just the right speed that keeps up with the Moon's rate of travel around its orbit, always keeping the same face toward us. The Moon used to spin much faster than this. Because of the Earth's tidal forces, the rock of the Moon bulges toward us and away from us just as the water on the Earth bulges toward and away from the Moon.

As the Moon was spinning on its axis faster than its rotation around its orbit, the elongated parts of the sphere were always trying to pull away from pointing exactly toward the Earth. But the bulge toward the Earth experienced bigger tidal forces than more distant parts, so the impact of those forces was to slow down the Moon's spin. This happened until the spin was at just the right speed to keep the same face toward us. Then the tidal forces were balanced out. This effect means that it is the norm for moons to keep one face toward their planet, rather than an oddity of our Moon.

Moons where this has happened are said to be in a synchronous orbit. A relativistic viewpoint should remind us that being on the Earth has no special privilege. The tidal force from the Moon is doing exactly the same thing to the Earth—it's just that the Moon's tidal force is a lot weaker, and the Earth a lot more massive, meaning there is a much smaller tendency to slow down our rotation. Even so, in the last billion years, the Earth has slowed down from an 18-hour day to a 24-hour one.

If the Earth lasted long enough, eventually the Moon would slow down the Earth's spin so that the Earth always had the same face to the Moon. Parts of the Earth would always have the Moon in the sky, parts would never see it. The Earth would still be rotating, but only once a year. In practice this will never happen, because our planet is likely to be destroyed when the Sun flares up to become a red giant in about 5 billion years' time, well before the Earth has had a chance to become synchronous.

Other moons feel even greater tidal forces. Jupiter's moon Io, for example, experiences around 250 times stronger forces than does the Moon as a result of the Earth.[13] This pulling and movement of the Jovian moon's matter produces a lot of friction. Although Jupiter's moons are very distant from the Sun, they may well have the best hope for finding extraterrestrial life in the solar system because of the heating that is in part produced by tidal forces. On Io, the result is dramatic volcanic eruptions.

Tidal forces also resulted in the discovery of a planet. The naked eye planets—Mercury, Venus, Mars, Jupiter, and Saturn—have been known since ancient times, but it was 1781 before another planet would be discovered. William Herschel, a German musician who had moved to England to better his career, was an amateur astronomer. He was searching out double stars with a homemade telescope outside his home in the city of Bath when he spotted what he first suspected to be a comet—a point of light that moved against the stars.[14]

After further observation, Herschel realized he had discovered a new planet, the first person ever to have this distinction. Herschel needed funds to support his work—the more he observed, the less time he had for performing music. The president of the Royal Society, Joseph Banks, suggested to Herschel that he could obtain financial backing from the English king if he named the new planet after him.

It seemed entirely possible that the run of outer planets would have been named Jupiter, Saturn, George . . . and this name (or more formally Georgium Sidus, George's star) was employed widely (along with planet Herschel) until the more consistent suggestion of Uranus made by German astronomer Johann Bode was universally accepted in 1860.

Uranus was discovered by visual scanning of the skies through a telescope, but the last of the major planets, Neptune, was predicted to exist before it was seen from effects of its tidal forces. In the nineteenth century, as measurements of planetary orbits became more accurate, it was noticed that the orbit of Uranus was being disturbed by something. Two astronomers, John C. Adams and Urbain Le Verrier, separately predicted that the cause of this disruption was tidal forces from an unknown planet. Neptune was discovered as the cause of those perturbations, in 1846.

If the tidal forces between planets can have a significant effect, it's nothing compared to what happens when stars come close together. We are so used to the Sun being an isolated body that it seems an unusual concept that two stars should orbit each other—in reality it is quite common. When Herschel discovered Uranus he was mapping double stars—stars that appear to be single to the naked eye, but turn out to be a pair of stars through a telescope. Many of these aren't binaries, just two stars that happen to be in a similar direction in the sky but that could be many light-years apart. But some are true binaries.

It is thought that around one third of stars in the Milky Way are binaries, so stars in pairs are anything but a rare oddity. When binary stars are relatively close together, the tidal forces each star can produce are considerable. Unlike planets, stars aren't solid, so this often results in plumes of stellar material being ripped from one star by the tidal forces and collected by the other as it forcefully eats its companion.

One of the dramatic possibilities of this gravitational effect is when a disk of new material (called an accretion disk) from a second star forms around a very hot white dwarf star. In this circumstance, the accumulation of material on the surface of the white dwarf can result in a nuclear chain reaction. Usually a star is a fusion-based nuclear reactor—here it becomes a catastrophic nuclear bomb.

The result is a nova. This is a contraction of the Latin words "stella nova"—new star. In the past, stars have suddenly appeared in the sky where they had not been seen before and were given the nova name. We now know that such sudden brightness is often the result of an explosion. The most dramatic are supernovas, where the majority of a star is destroyed, pushing out heavy elements, but in the form of nova seen in the white dwarf accretion disk, only the outer disk explodes, leaving the star to harvest more material from its neighbor and potentially go through the nova process over and over again.

Such interaction between stars is highly dramatic compared with anything we experience in or near the solar system (just as well for our comfort), yet in galactic terms it is small fry for the kind of phenomena that gravity can produce. Perhaps the most remarkable example of all is the quasar. Quasars (a contraction of quasi-stellar objects) appeared on first discovery to be distant stars, but they were like no other star. Their spectra, showing their composition, simply didn't fit with the way a star should look.[15]

A second worrying aspect of quasars was their radio output. Once radio astronomy was established, it was discovered that quasars were incredibly bright at radio frequencies—that's how they got spotted in the first place. The Dutch astronomer Martin Schmidt had the first clue as to the nature of quasars in the 1960s. He realized that the strange spectra made sense if the quasars were hugely redshifted.

When an object that emits light moves toward us, that light is blueshifted. The light gains extra energy, pushing it into the bluer end of the spectrum. Similarly objects moving away from us are redshifted. The American astronomer Edwin Hubble had established that almost every galaxy is moving away from ours as the universe expands, and the further away an object it is, the more it is redshifted. If Schmidt was right, the quasar he was working on (3C273) was the most distant object ever observed. Yet it was so bright it seemed like a star in our galaxy.

After much more study it was discovered that quasars are mysterious bodies that emit as much light as a typical galaxy from an area the size of the solar system, giving their unique appearance. Modern telescopes have showed that quasars tend to have a pair of "jets"—very energetic streams of glowing material, one spurting from either side of the quasar. So what is going on here?

A lot is still speculative, but the best suggestion is that a quasar is a young galaxy. Most galaxies are thought to have an extremely large black hole at their center. In a mature galaxy like our own Milky Way, this black hole will have cleared up the space around it, ridding it of debris. But in the young galaxy, that black hole would be pulling in vast quantities of nearby material.

As the matter plunges in toward the black hole under the vast gravitational pull of the massive body, it accelerates to a fair percentage of the speed of light, giving off huge amounts of radiation in the process (charged particles always give off light when they are accelerated). This is thought to be the source of the main quasar glow, a bright beacon at the heart of a young galaxy. As for the jets, there are still a number of possibilities.

One likely sounding suggestion is that the black hole is surrounded by a sphere of material kept in place by the spin of the body, like the material around the Sun that formed the solar system. This accretion disk would have large quantities of matter, except at the rotation poles, where material won't be spun in place. A pressure build up could then blast material out of the holes in the form of jets.

The gravitational pull of a planet not only has impressive tidal effects, but can also be used by spacecraft as a useful boost to their speed. This is a mechanism that was seen in an exaggerated fashion in the TV show *Star Trek*, where the *Enterprise* used the energy it gained in whipping around a star to travel in time. The reality is less dramatic—yet nonetheless surprising.

The approach is known as the slingshot effect, because the spacecraft making use of the effect experiences the kind of extra speed from being whipped around that a stone does when it is whirled around in a slingshot when compared with simply throwing the stone in a single action. (This is the old-fashioned kind of slingshot used by David against Goliath, rather than the modern version propelled by elastic.)

It might seem at first sight that there should be no net benefit. The idea is that the spacecraft heads toward a planet (or star), skims around it in an orbit, and then flies off in the opposite direction. It's obvious that as the spacecraft heads toward the planet it will get faster, accelerating with the gravitational pull. But then once it has passed around the planet and is heading away, the result should be a deceleration, slowing the spacecraft down. Why don't the two balance out, leaving it at the same speed it originally started at?

In a sense this does happen. The spacecraft does stay at the same speed—and yet at the same time it is speeded up. Let's take an example where the slingshot effect has actually been used, such as passing around Jupiter on the way out of the solar system. It's all a matter of relativity. We need to know what is being used as a comparison to measure speed against. As far as Jupiter is concerned, the probe will have left at the same speed as it entered its trajectory around the planet. But seen from the perspective of the solar system as a whole—the thing the probe is trying to escape—the spacecraft will have been speeded up.

This is all because Jupiter is itself moving around its orbit, at a tidy speed of around 13 kilometers per second. If the spacecraft heads toward Jupiter as Jupiter goes toward it, then the result will be to add Jupiter's speed to the craft. It would still be moving at the same speed with respect to Jupiter (though in a different direction), but much faster with respect to the solar system as a whole. In the process, Jupiter will have been slowed down in an exchange of energy, but by a negligible amount as it has so much more mass than the spacecraft.

A rather different effect of gravity and Newton's laws of motion applies in the case of the old puzzle of what would happen if all the occupants of China stood on chairs and jumped off at the same time. Would there be any impact on the Earth's orbit of such a massive impact as they came crashing to Earth? Could the effect even, as it has sometimes been suggested, be so great as to knock the Earth entirely out of its orbit?

There certainly would be an impact—this is as true with one person as a billion people in China, so to keep things simple let's look at what will happen if you personally decide to climb on a chair and jump off. There is a gravitational attraction between you and the Earth. You are attracted toward the Earth and the Earth is attracted toward you. However the Earth is so much more massive that to all intents and purposes we can say that you are the one being pulled to the ground.

In the short distance involved you won't come near the terminal velocity of a human body in air (the speed at which air resistance balances out the pull of gravity), which is around 120 miles per hour (54 meters per second). So you will accelerate by around 22 miles per hour (9.81 meters per second) each second of that brief fall, before hitting the Earth. Bang! How could this not knock the Earth out of its orbit? Very easily. If we look at a more meaningful example, we'll see why we aren't even looking at the right part of the exercise for a measurable effect.

Imagine you are in space, standing on a very small asteroid rather than on the Earth. You step up onto a chair that for convenience we have bolted to the asteroid. What happens? As you push down with your foot on the asteroid to step up, Newton's third law kicks in. If you are getting a push upward to move onto the seat of the chair, the asteroid must be getting a push down. So the result is that the asteroid moves downward as you move up. Then you jump off. You and the asteroid move toward each other—in this part of the process the asteroid is moving upward. Finally you hit each other. Assuming you don't bounce off, the energy of your movement goes into heat and noise (if there's air to carry it). Neither you nor the asteroid go anywhere.

Overall then, there should be a small movement of the Earth when you step up and another in the opposite direction as you fall down, but no movement caused by the impact. Net effect—nothing. Even if there had been an effect, it would be tiny. One billion people sounds like a lot and it is. Let's say everyone weighs 100 kilograms (220 pounds).

That makes for 10^{11} kilograms (10^{11} is 1 with 11 zeros after it) or 2.2×10^{11} pounds crashing into the Earth. But the Earth has a mass of 6×10^{24} kilograms (1.3×10^{25} pounds). It 60 trillion times more massive. Think of the impact of a dust mote hitting your body. The Earth simply wouldn't notice.

Considering how simple a concept jumping up under gravity

is, it can cause a surprising amount of confusion, to the extent that it was thought for some time that kangaroos somehow managed to defy gravity. It's all a matter of energy. Animals (us included) are able to move because of the energy they get from the food they consume. Any particular piece of food will contain a certain amount of energy (usually measured in calories), and the animal eating it can't use up more energy than that without eating into rapidly depleting bodily reserves. In practice it will only be able to access a percentage of the energy in the food, as no means of converting energy to work is 100 percent efficient.

When an animal moves it uses up energy—and jumping along like a kangaroo proves a considerable drain on that energy store, because there is both energy required to push the animal up into the air, overcoming gravity, and the effort needed to move along. When biologists compared the energy intake of the kangaroo with the energy required for all that jumping, it seemed to use more energy than it consumed, making the kangaroo defy the law of conservation of energy. If this were, true, kangaroos would make amazing power sources.

What was missed in these calculations is that the muscles of the kangaroo act like a bouncing rubber ball. When a rubber ball hits the ground, the ball deforms, absorbing energy from the collision. It then springs back into shape, transforming that briefly stored energy into kinetic energy of movement. It's like the energy stored in a spring or a stretched rubber band. Without any extra energy being put into the system, the ball jumps up in the air again. Similarly, the kangaroo's muscles store up energy as the animal hits the floor, enabling it to make subsequent bounces with less effort. It doesn't generate extra energy, it just stores up some of the energy of the collision, rather than wasting it all in heat. This is similar to the way electric cars transfer braking energy to their batteries, gaining extra charge.

Another example of a creature seeming to defy gravity that was once a popular topic in religious sermons is the bumblebee. Here we have an insect with a very large body compared with its tiny, fragile-looking wings. "Science has no answer as to why a bumblebee can fly," preachers would thunder. "The bumblebee defies gravity! Only God can make this happen!" Unlike the kangaroo, which really does appear to use more energy than it consumes, this is simply a myth. Bumblebee wings are certainly too small for it to glide like a plane or a soaring bird, but taking in the whole motion of the wings and the complex air movements involved, there is no problem. For the bumblebee it's a matter of aerodynamics and fluid flow, not an issue with gravity.

This doesn't mean that gravity doesn't have an influence on some modes of transport, though. Take floating along with the current on a river. Where are you headed? We are used to hearing that rivers run to the sea, but this is nonsense. It is perfectly possible for rivers to run away from the sea. All rivers do is to run downhill, thanks to the influence of gravity. Generally speaking, the sea is lower than most of the dry land, so the tendency is to run toward the sea, but it is entirely possible for there to be downhill runs ending in a lake that is far from the sea.

Whenever we head upward we are fighting the pull of gravity, but we can also use gravity to help us to travel. When a modern car is running down a hill in gear with no pressure on the accelerator pedal it uses no gasoline. The fuel injection system cuts out, leaving gravity alone to keep the car in motion. And in principle we could use gravity as a means to get anywhere on the planet—a total journey of around 42 minutes—without exerting any energy at all once the travel system is constructed.

Before you rush to buy shares in this amazing means of transport, it isn't a practical possibility. What it involves is digging a tunnel through the Earth. If we think of the extreme version of this, traveling to the exact opposite point on the surface of the Earth, involves covering a distance of around 12,500 kilometers (7,800 miles)—not bad for a 42-minute journey with no exertion of energy.

This feat assumes the tunnel has no air in it to provide resistance, but if this is the case, after jumping in (hopefully encased in a survival suit) you will be gradually accelerated until you flash through the Earth's center at around 7,900 meters per second (over 17,000 miles per hour). On the second half of the journey you will gradually slow down until you arrive at your destination at a standstill (where you will be quickly scooped up before you fall back to oscillate like a giant spring).

Rather neatly, the same time applies if you don't go to the exact opposite point. The distance will be less, but the acceleration will be less, too. In fact it should even work for a conventional tunnel through the Earth from any point to any other point. Provided it is straight, has no air in it, and you are prevented from making any contact with the sides of the tunnel, you should in theory fall through it in 42 minutes.

In practice, constructing a tunnel through the center of the Earth is unworkable. Before getting far into the crust things get too hot to handle, and even if the planet were solid all the way through, the engineering feat of drilling through 12,500 kilometers (7,800 miles) of rock (the longest tunnel in the world is nearly 100 times shorter) verges on the impossible. Yet the theory is still a very nice example of gravity at work.

We tend to think of the Earth's gravitational pull as uniform, the same around the planet, but this clearly isn't true. Even if the Earth were entirely uniform, made of the same stuff from surface to core, the gravitational pull would vary as we climbed mountains or traveled down canyons, because the distance to the center of the Earth would be varying and with it the gravitational pull. To make things harder still, the Earth bulges around the equator due to its spin. But it's also true that the Earth is anything but uniform.

This has been demonstrated powerfully since 2009 when one of the strangest scientific satellites in existence was launched.[16] The Gravity Field and Steady-State Ocean Circulation Explorer (shortened with some imagination to Goce) contains three delicately balanced pairs of platinum blocks that are used as accelerometers to measure variations in gravitational pull from the Earth's surface. As these changes are very small, the satellite needs to fly low—so low, that it can't maintain an orbit without help. Goce is one of the few satellites that has its own ion thruster engine to keep its orbit from deteriorating. In effect, this satellite is a drone spacecraft.

The first results from Goce, published in March 2011, are primarily aimed at getting a better understanding of ocean currents through the distribution of gravity on the Earth, but the satellite can also identify the tectonic boundaries that cause earthquakes. It will also give a better understanding of the physics of the Earth's interior and a global reference for measuring heights. Publicity in 2011 said that the probe showed the Earth was shaped like a potato. This was a highly misleading spin on the facts; the "potato" is a 3D map of the different gravitational pull levels, not the actual shape of the Earth. More importantly, the data is magnified by 10,000 to produce a visible effect—the reality is much more subtle.

The fact remains though that the acceleration due to gravity on the Earth is not a constant. It's something that influences the sea levels around the Earth, and that should have been borne in mind when the Channel tunnel was dug between England and France and ended up half a meter out on meeting up because the two countries had different definitions of sea level. The value for g, the acceleration due to gravity on Earth, is usually given as 9.81 meters (32 feet) per second per second, but Goce has mapped out the variations as the force shifts between 9.78 at the equator and 9.83 at the poles.

These examples of the way gravity can surprise us are entertaining, but are still eclipsed by the essence of gravity as an action at a distance. As long we consider gravitation an attraction, like Newton, we are left struggling to explain how it can possibly happen. This remained the case all the way through the twentieth century. But Albert Einstein did away with the need for such concerns, turning gravity into a kind of geometry. He would transform the deep question from "How does gravity act to provide a force at a distance?" to "How does matter warp space and time?" Gravity was about to receive a dose of relativity.

CHAPTER SIX

WARPING THE UNIVERSE

> *Today scientists describe the universe in terms of two basic partial theories—the general theory of relativity and quantum mechanics. They are the great intellectual achievements of the first half of the century. The general theory of relativity describes the force of gravity and the large-scale structure of the universe . . .*
>
> —*A Brief History of Time* (1988)
>
> Stephen Hawking

If you ask someone in the street what Einstein is famous for, they will say without hesitation $E=mc^2$. And also for being a weird old guy with a funny German accent and a shock of stand-up white hair. After a little thought they might also throw in relativity, by which they would mean special relativity, the strange implications of assuming that light travels at the same speed, however we move. Yet any physicist (and Einstein himself) would identify general relativity as the crowning glory of Einstein's career.

Einstein's life is so frequently covered that it seems almost unnecessary to describe it, and yet like Newton and Galileo before him, there is no doubt that his environment and upbringing had as much impact on his iconoclastic approach as did genetics. He was born in the southern German city of Ulm on March 14, 1879, into a family that certainly respected the work ethic, but never quite made it to success.[1]

Einstein's father Hermann had a string of businesses, partly funded by Einstein's mother Pauline's more wealthy family, but none could be described a great success. There is no doubt that Einstein senior put plenty of effort in, but he lacked a flair for business, or the luck to be in the right place at the right time. By contrast, Albert seemed to succeed despite not always putting in a lot of effort. He was stubborn, determined to do things his own way, and capable of looking at things in a totally revolutionary fashion. Einstein didn't so much think outside the box, as totally redefine it.

When Einstein was young, this innovative streak was anything but helpful. Although life at home with sister Maria (always known to Albert as Maja) and his parents seems to have been happy despite financial worries, school days for Einstein were fraught with difficulties because of his inability to accept matters, just because he was told that was the way things were. The German education system at the time was highly inflexible—it expected conformity from its pupils, but Einstein wanted to do things his own way at an age when most children would be happy to do as they were told.

By the time Einstein started school the family had moved from Ulm to Munich. Although his family was Jewish, they were nonpracticing and were happy to send young Albert to the local

Catholic school. Albert was less comfortable with the decision, not because of the religious education—which he largely ignored as he would very politely ignore organized religion all his life—but because of the school's rigidity of approach to education. It was here that the headmaster made one of those foot-in-mouth predictions that become famous and live on long after the man is forgotten. Albert Einstein, he believed, would never make a success of anything.

As a teenager, things got even worse. Einstein's next school, Luitpold Gymnasium, took an approach to education that stressed the classics and the humanities. Einstein struggled with the dead languages and was bored stiff by the rest. He was regarded as lazy and uncooperative, and there was no doubt a lot of truth in this. Even Einstein would probably have agreed with their assessment. He considered the education he got from books and from friends of the family, like the young scientist Max Talmud, to be much more significant than anything he learned at school. But Einstein was about to lose his anchor. His family moved to Italy.

Left behind to continue his education as his father made yet another dire attempt to succeed in business, Einstein gave up attending school just before he was expelled. He knew that he was soon to be signed up for compulsory military service, a thought that appalled him even more than the possibility of carrying on with his infuriatingly irrelevant education. At just sixteen he felt he could no longer remain a German citizen. His consistent trait of iron determination to do things his own way was about to change his life.

Italy with his parents did not seem an attractive alternative to Germany; instead Einstein settled on Switzerland. As well as having a German-speaking region, Switzerland seemed more forward looking, less set in its ways than his own country. What's more, in Zurich was the Federal Technology Institute (Eidgenössische Technische Hochschule or ETH for short), a university that specialized in science and technology, leaving behind those hated humanities and classics.

At least, that's what Einstein thought. While it was true the ETH was an excellent science school, it expected its students to be well rounded and the entrance examinations included the humanities. Einstein failed to get in. But his science results showed potential, and the ETH principal suggested he spend a year in a Swiss high school to get polished up for entry, then reapply. Einstein was younger than the usual students, so he had nothing to lose. The tactic worked. In 1896 Einstein reapplied and won a place.

There is no doubt that the ETH was a better environment for Einstein than his German schools, yet even here he didn't exactly shine. Like Newton before him on his undergraduate course, he was just too much of an original thinker to be considered one of the best students. Einstein did graduate—but only with a lot of help from a friend Marcel Grossman, who attended the lectures Einstein couldn't be bothered to go to and provided him with copious notes for his last-minute studying. As Einstein's physics professor commented, "You are a very clever boy, but you have one big fault: you will never allow yourself to be told anything."

As many people find, perhaps the most important experience of being at university for Einstein was a transformed social life. He was never one to spend all his time locked away with books and after a good number of girlfriends, he settled down to the serious pursuit of fellow student Mileva Maric. They eventually became engaged, and by the time he was 26 they were married and had had two children. The first, a girl called Lieserl, was born before the marriage and given away for adoption. The second, Hans Albert, was part of their new life.

Albert and Mileva settled in Bern, Switzerland. After gaining his degree, and Swiss citizenship in 1901, Einstein looked for a job and chose a surprisingly dull occupation—as a patent officer. A family friend of Marcel Grossman had given him an introduction to the head of the patent office when a vacancy just happened to have opened up. The job was for a Patent Officer (second class), but because of his lack of experience, Einstein started the post as Patent Officer (third class).

Einstein found the role a surprisingly perfect match for his talents. He had always considered himself impractical, yet he found it easy to visualize the inventions being put forward in the patent applications. He could construct these devices and confections in his brain and mentally experiment with them, finding flaws and spotting excellence. Not only did he find the job easy, there was little time pressure, giving him the opportunity to work on his own ideas, and in 1905 this freedom was to pay off in a big way.

In this year he wrote three outstanding scientific papers.

One, on the photoelectric effect (a paper suggesting that photons of light were real particles, and that formed one of the foundations of quantum theory) won him the Nobel Prize. Another on Brownian motion, where the impact of water molecules causes pollen grains in suspension to jerk about, provided the first theoretical basis that made it seem likely that atoms and molecules were real entities. And the third established special relativity.

Called *On the Electrodynamics of Moving Bodies* (or, in the original German *Zur Elektrodynamik bewegter Körper*) the title of this paper gives little away, yet special relativity would transform physics, showing that Newton's laws of motion reflected a special case for bodies that were not moving particularly quickly. Special relativity applies one extra piece of information to Newton's laws, a piece of information that has profound consequences.[2]

Einstein was aware that the Scottish physicist James Clerk Maxwell had explained the nature of light at the end of the previous century. Light, Maxwell discovered, was an interaction of electricity and magnetism. Move a source of electricity and you produced magnetism. Move a magnet and you made electricity. If you got a wave of electricity moving at just the right speed it would produce magnetism which itself produced electricity, which produced magnetism and so on. But this alternating dance could only happen at one speed in any particular medium. In a vacuum this speed was around 300,000 kilometers (186,000 miles) per second. The speed of light.

If the waves were traveling at any other speed, this interacting dance could not take place. The whole thing would collapse. But

Galileo had shown in a concept called relativity that nothing had an absolute speed. If a car is traveling at 50 miles per hour and you drive alongside it in another car at 50 miles per hour, the first car isn't moving for you. Similarly we see the Earth as fixed, unmoving, even though it is hurtling around its orbit at over 60,000 miles per hour, because we, too, are moving at the same speed.

If the same basic relativity applied to light, then anything moving alongside a sunbeam at the speed of light, would see the light as stopped. But if that were the case, the light would no longer exist. In practice any movement with respect to a light beam should make it disappear, because for that moving observer the light would not be moving at the right speed. As this doesn't happen, Einstein was left with one startling conclusion; that, unlike anything else, light is not influenced by relativity. It continues at the same speed; however, we move with respect to it.

The result of pinning down this one value is to free up others. We pay a big but necessary price for fixing the speed of light. Specifically anything (other than light) that is moving will change compared with its properties when stationary. Time on a moving object slows down. The object gets shorter in the direction of motion. And its mass increases. These are all real effects that were predicted by Einstein's version of relativity dealing with steady motion, which would be given the name special relativity. ("Special" in the sense that it was a special case, not because it is remarkable.)

Special relativity was also responsible for that most famous of formulas, $E=mc^2$. The derivation of this simple formula linking energy, mass, and the speed of light (c) is not trivial, but it is a consequence of the changes brought about in our understanding of mass and the energy of movement produced by special relativity. We know that there is a certain amount of energy in a moving body due to its movement—its kinetic energy. If you doubt this, you have never played dodgeball.

Combine the concept of kinetic energy with the change in mass that special relativity predicts when an object moves and you have a new formulation that introduces a relationship between mass and energy, a relationship that is played out in that iconic formula $E = mc^2$.

With these three outstanding papers under his belt, Einstein was soon to leave the patent office behind. By 1908 his fame had spread and academia called. But he was to have another simple yet profound thought while still at the patent office. Sitting in that boringly ordinary office in Bern, a thought came to Einstein that would lead to general relativity and a radical transformation of our understanding of gravity.

All too often, breakthrough moments seem to be fictional. We tend to create them from nothing, out of the need that human beings have to hear about heroes and challenges overcome in a flash of sudden inspiration. Real scientific endeavors often involve long, collaborative processes without a single clear example of a "Eureka!" moment. Yet scientists do sometimes claim that they really occur.

Some cynical historians of science suggest that these light

bulb moments are simplifications after the fact, a fake memory that the scientist themselves may well believe in, but a memory of an event that never really occurred, a back formation from all their later work. But I am inclined to go with the scientists' own view. Ideas do come suddenly and fully formed, even in such a complex field.

I do believe that Newton had that moment of inspiration while looking at an apple. And that Einstein's general relativity was born from an idle thought that he had in the Swiss patent office in 1907. Here are Einstein's own words: "I was sitting in a chair in the patent office at Bern when all of a sudden a thought occurred to me: 'If a person falls freely he will not feel his own weight.' I was startled. The simple thought made a deep impression on me. It impelled me toward a theory of gravitation."[3] He would describe this as "The happiest thought of my life."[4]

What Einstein's thought suggested is now known as the principle of equivalence. This says that the effect of gravity and the effect of acceleration are identical—the two phenomena are equivalent. Imagine being in an enclosed spaceship that is perfectly proofed against sound and light, with no vibrations or other detectable movements. You are standing on the "floor" of the ship, feeling your normal weight. The principle of equivalence says that you have no way of knowing if the ship is stationary on the Earth, and you are feeling the pull of gravity, or if the ship is accelerating at the same speed as the acceleration due to gravity—around 9.8 meters (32 feet) per second per second, pushing you to the floor with its acceleration.

In both cases you would be pressed against the floor of the ship with your normal weight. No experiments you could do would tell you whether you were feeling the influence of gravity or accelerating. At this stage, Einstein's idea is very much the equivalent of classical relativity. When Galileo first came up with the idea of relativity he imagined being on an enclosed boat, moving at a steady speed, and said there was nothing you could do to discover whether or not the boat was moving. But general relativity goes beyond classical relativity by dealing with acceleration, and as a result begins to give us some insights into the nature of gravity.

Before going any further I ought to clear up one frequently misunderstood aspect of the principle of equivalence. People like proving Einstein wrong. There's something of an industry devoted to this, and I've lost count of the number of books I have been sent with some novel physics theory that shows that the great man was mistaken. Occasionally such books will say that the principle of equivalence is wrong. In a sense they're correct—but it is in a very mean-spirited sense that is irrelevant to the importance of general relativity.

Let's go back to that imaginary spaceship. In reality, if you were in that enclosed ship, and you had the right equipment on board, you *could* tell whether you were feeling the pull of gravity or were accelerating. Let's say you measured the strength of the apparent gravity right up in the nose of the spaceship and way down the back by the engines. If the ship were accelerating, the apparent "gravitational pull" would be the same at both ends of the ship. But if the spaceship stood still on the Earth, the gravitational pull would be slightly weaker in the nose, because it is a little further away from the center of the Earth.

Of course, Galileo could also have found out he was moving on his enclosed ship if he had the right equipment (GPS for example), but the difference between the two examples is that the basic laws of physics would operate exactly the same whether or not the ship was moving in Galileo's case. Galileo could only tell the difference using technology that provided a way to look outside the ship. On Einstein's spacecraft there would be a noticeable difference in the force of gravity at different ends of the ship in one case, but not in the other, the exact measure that is supposed to be equivalent.

Although sophisticated instruments are needed to tell the difference between being on the surface of the Earth and under 1g of acceleration, it's even easier to make the equivalent distinction between being in free fall under the pull of gravity and being weightless with no gravitational pull. Imagine you had four stones and let go of them at the same time. Two stones are on one side of the ship and two on the other, each pair of stones positioned one above each other. If you are floating weightless in space with no gravitational pull, then the stones will just stay in place forever. But it would be different if your spaceship was in free fall toward a planet.

In free fall, the stones that are placed above one another will gradually get further apart. The lower stone will feel a slightly stronger gravitational pull, so it will accelerate slightly faster than the higher stone. What's more, the stones on the two sides of the cabin would be pulled toward each other because they are not heading downward on parallel lines, but on converging lines, heading for the center of the body you are freefalling toward. The principle of equivalence doesn't really work across a wide open space.

However, using this apparent flaw in the principle of equivalence to attack Einstein and his theory misses the point. In fact a point is exactly what is needed to do away with the problem. The principle of equivalence doesn't have to work at every location on a spacecraft simultaneously—the idea of making measurements onboard is just a nice picture to make equivalence clearer. The principle only needs to operate at any single point in space, a single location within the spaceship at which to make the comparison, and in those circumstances it works perfectly.

Let's take a look at Einstein's specific statement of his original idea. It wasn't about comparing being on Earth to accelerating; it was about the apparent zero gravity of free fall. "If a person falls freely he will not feel his own weight." We have been assuming that this is the same as the equivalence of gravity and acceleration, but it's worth making sure. Take, for example, the International Space Station (ISS). You have probably seen TV pictures of the ISS astronauts floating around pretty well weightless. They are, of course further away from the Earth than we are, so they feel less gravitational pull than we do on the ground. They should weigh less. And they do.

Let's risk a brief calculation. Remember the formula for the force of gravity:

$$F=Gm_1m_2/r^2$$

We can use this to work out the difference in the force of gravity that the astronauts feel on the ISS from the level we experience on the ground. Luckily for those of us with low tolerance for math, practically everything cancels out. G is the same, m_1 (the mass of the Earth) is the same and m_2 (the mass of the person) is the same. So the ratio of the gravitational forces $Force_{ISS}/Force_{Earth}$ is just r^2_{Earth}/r^2_{ISS}.

Now here's the thing. The ISS flies at an average of 350 kilometers (217 miles) over the Earth's surface. But the radius of the Earth is around 6,370 kilometers (3,958 miles). So r_{ISS} is really not very different from r_{Earth}. It's just $r_{Earth}+350$, or 6,720 kilometers. So the ratio of the forces is (6,370×6,370)/(6,720×6,720)—which works out around 0.9. The force of gravity on the ISS is 90 percent that on the Earth's surface.

And this would be exactly what the astronauts would feel if they were sitting 350 kilometers up, on top of a high tower. But they aren't, they are in orbit. As Robert Hooke pointed out back in the seventeenth century, when a satellite like the ISS is in orbit, it is falling. Falling freely under the force of gravity. The only reason it doesn't crash to the Earth is that it keeps missing. Because as well as falling it is moving sideways at a tangent to the Earth.

If it only had this sideways motion it would fly off into space. But the combination of the two movements keeps it circling around the Earth, always falling, always missing.

Here is where we see Einstein's picture in action. Because the astronauts are falling they don't feel their weight. The acceleration due to gravity exactly cancels out their weight, generated by the 90 percent of the force of gravity felt on the surface. They float because the two are equivalent. We have the pull of gravity and the acceleration and the two cancel each other out, something that is only possible if they are equivalent forces. The only difference with the comparison of being on Earth or in an accelerating spacecraft is that the floating ISS astronaut both accelerates and feels the pull of gravity at the same time, and the two cancel out.

Once Einstein had accepted that the effect of gravity and acceleration are equivalent, he was able to take another small step that would result in a giant leap in mankind's understanding of the nature of space, time, and gravity. Imagine we are back in our enclosed spaceship, and just for the moment, let's assume it truly is accelerating under power. Now let's emulate the beginnings of one of Newton's most famous experiments.

When Newton was investigating the nature of light and color he did so with a prism, a toy bought at a local fair. To try out the prism he pulled down the blinds in his room, then made a pinhole in the blinds. (Newton was never one to worry too much about university property—he later accidentally burned down the shack he had built onto the college building for his experiments, risking sending Trinity College's beautiful Old Court up in flames.) When Newton made the hole, a narrow beam of light shot across the room, hit the prism, and made a rainbow.

In our spaceship we will also make a narrow opening to let in some sunlight. A pinhole wouldn't be a good move in the vacuum of space, so we will just take the cover off a tiny porthole. As with Newton, a beam of light shoots across the room. But here there is a difference. The spaceship is moving. It is accelerating at right angles to the beam of light. In the time it takes the light to cross the room, the ship will have moved. If we had filled the room with smoke we would see a slightly curved beam of light, tracing a parabola that bends downward.

Of course the effect would not be large. The light is likely to be moving much faster than the ship. At any speed we can achieve currently the curvature of the light beam would not be noticeable to the eye. Nonetheless, with suitable equipment the effect could be measurable. The light ray would inevitably bend as it passed through the ship.

Now let's return to our principle of equivalence and assume I told you a little fib. Inside the ship, you can't tell whether you are accelerating or you are stationary under the influence of gravity. I told you that the ship was accelerating, but this was a lie. It is in fact sitting on the ground, not moving, feeling the pull of gravity. Yet equivalence tells us that you can't do an experiment that will distinguish between the two. So the light will still bend as it crosses the room, by just the same amount.

Here comes the giant leap. Rather than just think, "Hmm, light is more like matter than we thought, it is attracted by gravity," which seems a perfectly reasonably way of interpreting the results, Einstein looked at the thought experiment in a different way. It's a bit like one of those pictures where you can see a cube in two orientations—suddenly the image flips. Einstein's picture of what was happening in these circumstances undertook such a flip.

All the thinking Einstein had already done about relativity must have helped. There were two key starting points to special relativity. The first was the importance of looking at things from different viewpoints, and the second was the realization that light went its own merry way regardless of what was happening around it. If Einstein took the viewpoint of the light itself, would it not see what was happening under the pull of gravity differently? What if the light was still traveling in a straight line from its own viewpoint? There was a way for this to be possible—if space itself was twisted. Or, more accurately if space-time was warped.

To grasp the enormous consequences of Einstein's inspiration we need to take a step back for a moment and think about the nature of space and time. In our usual conception, space and time are very different things. Space is something we move around in at will. I can move my hand left and right, up and down, back and forth, in any combination of the three spatial dimensions. Time seems very different. Time is something that ticks along in a particular direction. I can't control how my hand travels through time, if it can be said to travel at all.

However Einstein was not the first person to see the potential for a tie-up of space and time. In his 1895 novel *The Time Ma-*

chine, H. G. Wells had suggested that time was indeed a fourth dimension:[5]

> "Clearly," the Time Traveler proceeded, "any real body must have extension in four directions: it must have Length, Breadth, Thickness and—Duration. . . . There are really four dimensions, three which we call the three planes of Space and, a fourth, Time. There is, however, a tendency to draw an unreal distinction between the former three dimensions and the latter . . ."

Once Einstein developed special relativity, it became clear that space and time were closely linked. Simply by moving through space it was possible to influence the flow of time as witnessed by an outside observer. It was impossible to think purely about an effect in space without also considering its impact on time. And general relativity would pull the two together in the tight mesh of space-time.

Space-time was not Einstein's baby. It was conceived by the Russian-born, but German-raised mathematician Hermann Minkowski. Within a couple of years of Einstein's publication of his paper on special relativity, Minkowski was exploring the implications of the theory in the context of a four-dimensional entity called space-time that treated time as a dimension just as Wells had predicted.

Although involving four dimensions, each at right angles to the other, and so impossible to imagine literally, space-time is often portrayed as a two-dimensional or three-dimensional image, simply by ignoring one or two of the spatial dimensions. This isn't as bad as it seems. The extra dimensions in space aren't needed to understand what's going on, as the spatial dimensions are interchangeable. The way we choose the axes for the three dimensions of space is arbitrary. With a single object moving I can always orient those axes so the movement is in a single spatial dimension.

So a simple Minkowski diagram in two dimensions portraying space and time would be something like the example below.

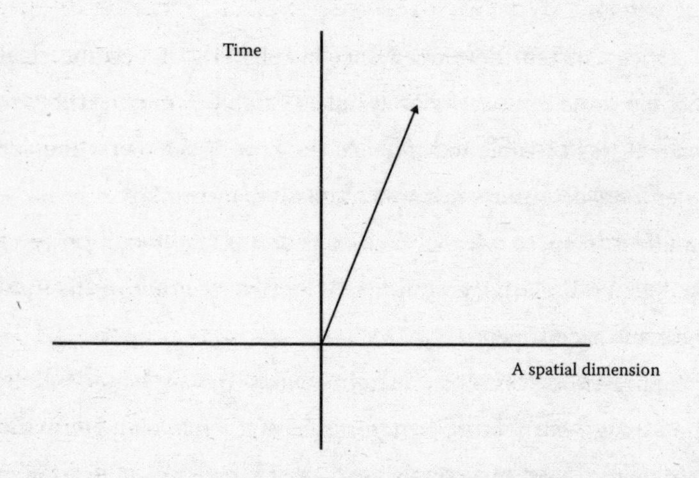

In this very simple Minkowski space-time diagram, time runs up the page and space across. The arrow shows a spaceship traveling at constant speed. As it moves forward in time (up the diagram) it travels through space (to the right). A stationary ship would travel straight up the time axis.

In a Minkowski diagram it becomes much easier to understand what is happening with various applications of special relativity. But it also gives us a picture of space-time that will be crucial for general relativity, because space-time is the very fabric of reality that this diagram represents a slice through, and it is the whole of four-dimensional space-time that is warped by an object's mass to produce the effects of general relativity.

It should be stressed, though, that space-time is not just a convenient way to look at relativity; it is a fundamental requirement to understand what is going on, because of the way special relativity throws our established and "natural" viewpoint out of the window. To be able to do science at all we rely on the idea that things don't change in an arbitrary fashion. If, for example, I do an experiment in two different places, I expect the experiment to work the same way. It is this consistency that often distinguishes between science and pseudoscience.

Of course it is entirely possible that an experiment will work out differently in different locations. If, for example, I do an experiment on the surface of the Earth and on the "surface" of the Sun I would expect some different results due to the differences in gravitational pull and in temperature. (Not to mention the lack of a breathable atmosphere.) It wouldn't exactly be difficult to notice a change in outcomes. But if the only difference is position, or time, or whether or not the experiment is in steady motion (remember Galileo) we would traditionally have expected the experiment to come up with the same results. This is a principle scientists refer to as invariance.

Unfortunately, special relativity threw invariance in space or invariance in time out of the window. Just moving relative to something else was enough to distort space or to shift time. But what Minkowski demonstrated—making his confection so valuable—is that unlike space alone, "distances" in space-time are invariant even under the most extreme provocation that special relativity can throw at your experiments. Any shift in one dimension is balanced out by the other.[6]

The only trouble with this assertion is that it is almost impossible to get our brains into the right mode for dealing with space-time. As soon as I mention a distance in space-time our mind inevitably latches on to spatial dimensions. Distances are traditionally a measure of space alone. But here, when we say that distance is invariant in Minkowski's space-time, we mean a measure of how far away two points are from each other in time and space.

Let's look at that simple diagram again but add a little more detail.

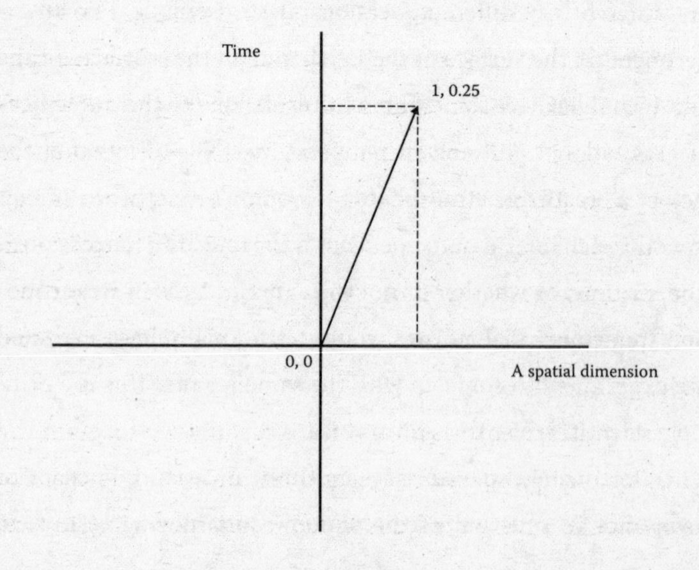

Now we are defining two points along the journey of the craft in space-time. We have arbitrarily started at 0 on the time axis and 0 on the space axis. We can do this wherever and whenever we like—all measurements have to be relative to something. We might think that "really" that start point was noon on May 15, 2102, in time, and at a spatial location we could identify by measurements with respect to the position of the center of the Earth at that time—but it's absolutely arbitrary.

Using the same starting location, the second point in space-time is 1 on the time axis and 0.25 on the space axis. Here it makes sense to adopt a simple convention of measuring time in years and distances in light-years (a light-year is just the distance light travels through space in a year, about 9,460 billion kilometers [5,878 billion miles]). So these numbers tell us that the second point, the head of the arrow, is 1 year along the time axis from the original point and 0.25 light-years along the space axis.

From this we can see that the spaceship is traveling at a quarter of the speed of light. It isn't accelerating, otherwise the path it took would not be a straight line. There is more that comes across here. Because of the units we've chosen, a beam of light would head out at exactly 45 degrees between the time axis and the space axis. Anything physical, limited by relativity to the speed of light, could be expected to take place above that line of light.

Although a Minkowski diagram is just that—a diagram that gives us a mental picture of what space-time is—it is useful to keep in mind when we talk about general relativity's implications of warping space-time. By imagining something twisting the sheet of paper a Minkowski diagram is printed on we can get an image of what is taking place. The straight line of a beam of light, unwaveringly shooting along at 45 degrees would now appear to twist into a curve.

As we aren't really concerned with special relativity, I don't want to spend too long on the diagram, but one important thing it is worth mentioning that is absolutely fundamental to general relativity is that we mustn't assume normal geometry applies. Just because the Minkowski diagram looks like a traditional two-dimensional diagram of space, we shouldn't confuse it with one. It is a space *and time* diagram.

While it is true in space alone that we can work out the length of a diagonal using Pythagoras's theorem, this doesn't necessarily apply to time. When taking into account some essential assumptions about the universe that would eventually be plugged into general relativity—notably that the universe should be the same everywhere and that cause should precede effect for any two points that light can connect between the time of the cause and the time of the effect—it turns out that Minkowski space-time doesn't follow Pythagoras.

Pythagoras would tell us that to work out a distance in space-time, that useful invariant that the Minkowski diagram produces, we need to add the square of the distance to the square of the "time distance" and take the square root. The squared space-time distance would be $(c \times \text{time})^2 + \text{distance}^2$ (where c is the speed of light—we need this to transform time into a distance).

In Minkowski space-time, though, the combination of spatial distance and distance in time works out differently. Here it is $(c \times time)^2 - distance^2$. We take away the square of the spatial distance rather than add it. It is this combined distance that remains invariant. One of the effects of this crucial change of sign is that an object in Minkowski space can never stray into the region where time flows backward.

You might think that if space-time is warped by the presence of matter as Einstein thought, then this warping should be obvious in everyday life. Shouldn't we be able to see around corners, for example? But just because something is curved—very significantly curved—doesn't mean that you can notice it in day-to-day life. We live on a planet that is roughly spherical. That's not a subtle bit of curvature. A "straight line" on the surface of the Earth—the shortest distance between two points—is a curve called a great circle. The clue is in the name. On the Earth a straight line is not straight, it's a circle, and a circle is more than trivially curved.

We don't notice this (and so tend to be a bit surprised when onboard a plane the display shows that your route from A to B is along a curve) because the kind of distances over which the curvature is noticeable are outside our normal visual capabilities. In the best conditions you can see around 25 kilometers (around 15 miles); in a city, visibility may be only a few hundred meters. You are not going to notice a curve in a sphere with a circumference of around 40,000 kilometers (24,860 miles).

Similarly we don't see light warping around corners in our

local area because the effect is relatively small. It's not that we can't see the impact of warps in space-time—they are all around us. But for the specific effect of light being bent into a visible curve we need some very concentrated mass. As we'll see, this does happen—in fact it's the way that general relativity was first tested—but it isn't going to happen visibly to a laser fired near the corner of a building.

Even so, Einstein's warp in the very fabric of reality is with us all the time in the apparent force of gravity. Space-time is constantly being warped and twisted, as each bit of mass in the universe acts to change the shape of its environment. General relativity says that force of gravity is just a tendency of bodies to move along the warps in space-time caused by the distortion of the fabric of reality by mass.

According to general relativity, gravitation is not so much one mass acting on another as mass acting on the intervening space-time, which impacts on the other mass—this way the whole problem with Newton's vision of gravity goes away. With general relativity in charge, gravitation is no longer an action at a distance. In a sense we return to the picture of there being a kind of aether through which the effects of gravity are transmitted—we just happen to know that this ether is space-time itself.

It might seem a big leap to go from a curve in space-time to gravity as we are used to it. Although Newton's picture requires the strange action at a distance that is attraction, the outcome seems entirely reasonable. If such an attraction exists, then two objects should pull toward each other. It's less obvious why bodies should be attracted and planets orbit with general relativity's curved space-time.

The British physicists Brian Cox and Jeff Forshaw have an interesting analogy for gravity as curved space. Imagine two people on the surface of the Earth, heading off north from the equator. They leave the equator 100 kilometers (62 miles) apart on parallel lines. Yet by the time they reach the North Pole, they will bump into each other—in effect they will have been attracted together by the curvature of the Earth.[7]

General relativity's attraction is a bit like this. A curvature in space-time means that two moving objects that are seemingly unconnected can be attracted toward each other. Which is all very well for things that are moving, but it doesn't explain why two objects are attracted together if they aren't moving. We need a bit more explanation for two stationary bodies (at least, stationary with respect to each other) that are attracted toward each other by gravitation. Enter the rubber sheet.

Read pretty well any book that mentions general relativity and you will come across the image of a massive object distorting a rubber sheet. I have not been able to pin down where this analogy originated, though it seems widely agreed that the first picture of the expanding universe being like the surface of a balloon came from that great exponent of general relativity, the British astrophysicist Sir Arthur Eddington—and there could be an inspirational link between the two concepts.[8]

In this popular model, the rubber sheet is space-time. When an object with mass (I hesitate to say "massive object," because "massive" has come to mean extremely large, but anything that has mass is a massive object) is placed in the rubber sheet of space-time, the result is to distort the sheet, stretching it so that the object sits at the bottom of a well of stretched rubber—the more mass that is crammed into the object, the deeper the well.

Now imagine a stream of photons coming along—a ray of light from a distant star, say. The photons are passing through space-time in a straight line. We can imagine the light as a colored thread running through the rubber sheet. When that thread encounters the dip in the sheet, even though those photons continue from moment to moment to travel in a straight line within space-time, they will be moving through a warped sheet. The result is that the photons will bend around the object. The gravitational pull of the object will have distorted the path of light.

The same goes for the way gravity will influence a passing object with mass—it will still experience a warp in space-time and will be bent off its straight course. So a planet orbiting the Sun, for example, is really traveling along in a good Newtonian straight line. There is no force at a distance. There is nothing pulling it off its straight line path. Instead the medium through which it is moving is being warped, so the straight line becomes a curved orbit.

So far, so good. But I have also frequently seen the same model used to explain basic gravitational attraction, and initially it is not clear how the rubber sheet model can explain why stationary objects should start moving. Imagine that I am somehow hovering in space near a massive object, not moving with respect to it. We'll call the massive object "the Earth." I have another object with mass in my hand—let's say, for Newtonian elegance, an apple. I let the apple go. What happens?

As Newton knew so well, the apple falls. Those who treat the rubber sheet model in a cavalier fashion say: "Yes, well, the apple would fall. Imagine you had a big heavy ball on a rubber sheet, causing a deep depression in the sheet. You put a small ball bearing near the edge of the depression and let go. The ball bearing rolls down the slope until it hits the big heavy ball. Similarly, the apple will fall down the gravity well of the Earth and hit the ground."

There's a problem with this picture, though. What is it that makes the ball bearing roll down the rubber sheet in this version of the analogy? It's gravity. It wouldn't work in zero gravity. Even if the big ball can be said to produce a dip because it is embedded in the rubber sheet, under weightless conditions the ball bearing would just sit there in its own little gravity well, stubbornly ignoring the bigger dip produced by the big ball. The use of gravity to explain gravity makes this a circular argument. It has no value.

We've got a problem and there are two possible ways out. One is simply to accept that we have stretched the analogy too far. In fact, stretched it so far that the rubber sheet breaks. This argument says that the rubber sheet model was useful to show a trajectory is modified by the distortion of space-time by mass, but is irrelevant for our apple. There are plenty of examples of analogies being taken outside their comfort zones and falling apart.

However, there is a much better way to get around the problem, one that means the rubber sheet analogy can still be used to explain the descent of the apple. It's a better solution, but one that is not obvious, as I discovered when I began asking physics professors if there was any way to use the rubber sheet model in this case.

The majority of academics I approached for a suggestion didn't reply. This might seem unsurprising, but on the whole professional scientists are happy to answer meaningful queries. The apparent implication here is that either they thought there was no way to apply the model (back to the overstretched analogy), or they didn't understand how it could be applied. (To be fair, few of the people I approached were experts in general relativity, though they were all physicists.)[9]

Where I did get replies, they tended to be along the lines of either irrelevance or that the mathematics works, so let's not worry about the analogy. A typical argument from irrelevance was: "A force is a force, of course, and asking why/how it causes an acceleration already seems a bit circular." However asking how you can have a force producing action at a distance is not at all circular, but something that has worried people for a long time. There still needs to be a cause.

The alternative approach of not worrying about what's happening as long as the mathematics works has a strong history in science, particularly in physics. Matrix mechanics, one of the early components of quantum mechanics, was based entirely on a mathematical concept that had no analog to explain its link to reality. There was no analogy, no mental picture that could be used to explain what was going on. The numbers worked without any good way to describe how they modeled what was observed.

I had an explanation of the rubber sheet for stationary objects that was close to this "just stick to the math" view initially from one professor—or so it seemed. I was told "As you correctly say, in the 'bowling ball on a rubber sheet' picture, the sheet represents space-time. In space-time the 4-velocity of a particle is just the tangent unit-4-vector of the path traced by the particle. Since a curve in the sheet at a certain point has always a tangent 4-vector, the particle has to follow the path with the corresponding 3-velocity."

This was useful to know, I'm sure, but didn't exactly illuminate things. But I'm pleased to say, after a little probing, that same professor (Friedrich Wilhelm Hehl of the University of Cologne) translated this explanation into a more amenable form. When we think of a photon traveling through the rubber sheet, the natural inclination is to imagine three-dimensional space being distorted by the planet (or whatever the object of mass is). But this picture is wrong.

General relativity does not say that massive objects distort space. It tells us that there is a distortion in *space-time*. When we think of space being distorted we are suffering from the same illusion as we were when trying to identify an invariant distance in Minkowski space-time.

Once we realize the distortion is in space-time, the explanation for the apple's fall becomes apparent. If just space were distorted by the Earth's mass, there is no reason for the apple to start moving. But when we consider space-time, the apple doesn't start off stationary. It is already moving—through time. If space-time is distorted by the Earth two things happen. One is that the rubber sheet of space-time warps. What was previously a movement in time alone shifts into a different direction within space-time, so it now is a movement in both space and time. The object has to start to move.

If this were just a kink in a flexible but non-stretchy space-time, the result would be that as movement in space increased, movement in time decreased. We would expect movement through time to slow down. And general relativity tells us that it should. Gravity slows down time. But the distortion of space-time is a stretch as well as a warp, so we can't necessarily expect a one-to-one correspondence between the changes in the movement in space and time.

Overall, the rubber sheet model holds up remarkably well, provided we remember to distort space-time and not space alone, something that is necessary for general relativity to work as a theory, but that we (arguably even some physics professors) have difficulty getting our brains around.

It's handy when exploring the effects of general relativity to consider the impact on light, because a photon is the ultimate moving object. Light has to move to exist, and always moves at the same speed in any particular medium, so we don't have to worry about acceleration and can concentrate on the basics. Unless it interacts with matter, light always goes in a straight line. Period. There is no arguing with this.

You might say, "Yes, but light doesn't go in a straight line when it hits a mirror or passes through a lens," but that is an interaction with matter. A photon comes in, hits the mirror, is absorbed by an atom in the mirror and is reemitted to fly off in a different direction. That is the essence of quantum electrodynamics, the science of the interaction of light and matter. But light is also influenced by gravity, as we have seen, because space-time is warped by bodies of matter.

Let's go back briefly to a scene we met earlier. We fired a bullet horizontally and at the same time dropped a bullet. The bullet fired out of the gun and hit the ground at the same time as the dropped bullet. But with general relativity in place we can take that comparison further. Because gravity bends space-time, and that influences the path of light just the same way that it works on solid matter, if we drop a bullet at the same time as sending a laser light pulse out horizontally, the light pulse will also hit the ground at the same time as the bullet does.

It's easy to react by thinking: "Surely the bullet will fall faster than a weightless photon?" But here you are falling into the trap that Galileo did away with. The amount of weight involved is irrelevant to the speed something falls. And Einstein shows us that this is because gravitation is about curvature of space-time and doesn't care about mass when it comes to the speed of falling. The photons in the laser pulse will experience exactly the same warping as anything else.

The other reason we doubt the simultaneous fall is that we really can't cope with the magnitude of the speed of light. The dropped bullet might take around a second to hit the ground. In that time, light will have traveled 300,000 kilometers (186,000 miles). Long before it hit the ground it would have run out of Earth and be far out in space. You would need a huge (but very low density) planet to be able to carry out this experiment for real, as the surface has to be approximately level over such a vast distance.

Remarkably, Newton wondered if gravity could have this kind of effect on light. He had no conception of a warped space-time that would make such an observation likely, but in the third volume of his book on light, *Opticks,* Newton speculates on a number of possibilities in his "Queries" section.[10] Query number one asks whether bodies will act on light at a distance, bending its rays. Bear in mind that Newton thought light was made up of tiny particles called corpuscles, and was merely extending the same effect as keeps the planets in their place to suggest that light, too, would be susceptible to gravity's influence.

Light, as I've already mentioned, was used in the first experimental proof of general relativity. General relativity predicts that a body as massive as the Sun should have a measurable impact on light that is passing close by it, bending the light off its expected straight line trajectory. (Or more accurately, bending space-time so light's straight line path appears twisted.) So if you have light from a star in the sky traveling very close to the edge of the Sun as it comes toward Earth, the star should appear shifted slightly toward the Sun from its "actual" position.

That's all very well in principle, but we don't see stars in the daytime because of all the scattered light from the Sun—all the other stars are washed out. What's more, the last place you are going to see one is right up close to the Sun's edge. It seems unlikely that Einstein would ever get such direct proof of relativity. Unless a measurement could be made during a solar eclipse.

CHAPTER SEVEN

EINSTEIN'S MASTERPIECE

> *Einstein, my upset stomach hates your theory [of General Relativity]—it almost hates you yourself! How am I to provide for my students? What am I to answer to the philosophers?!!*
>
> Letter to Albert Einstein (November 20, 1919), Paul Ehrenfest, quoted in *Paul Ehrenfest: The Making of a Theoretical Physicist* (1970)—Martin J. Klein

Einstein first presented his theory of general relativity in 1915 at the Prussian Academy of Sciences in Berlin. Though the mathematics behind the theory is less accessible than special relativity, his new development was picked up on by the newspapers as a major story—the idea of reality itself being warped was fascinating—and the news media also had a significant interest in attempts to discover whether observations matched Einstein's elegant but mathematically challenging theory.

If everything had gone smoothly there could have been observations to back up general relativity at around the same time as the publication of the theory. Einstein had discussed the basic concepts as far back a 1912 with another German scientist, Erwin

Freundlich. It had then taken Einstein a good 3 years to get on top of the mathematics required to turn his theory into a detailed prediction. He struggled with the techniques required and had to get more help than he was comfortable with. But he was already far enough advanced by 1914 for Freundlich to set off an expedition to capture data that could reinforce the theory.[1]

Unfortunately for Einstein (and arguably even more so for Freundlich) things did not go well. Any particular total eclipse is only visible in a relatively narrow band of the Earth's surface, so making the measurements meant getting to a point on the Earth where the effects of the eclipse could be monitored. In practical terms this meant Freundlich heading off for the Crimea. But by August 1914, when the expedition took place, Germany was at war with Russia.

The Russians captured Freundlich and mistook his telescopes and other scientific equipment for spying gear. Freundlich was briefly held prisoner until the end of August, when he was returned to Germany as part of a prisoner exchange. There is no record of what happened to his equipment, but it seems unlikely he was allowed to take it home with him. More to the point, he returned empty handed as far as useful scientific data goes.

On average total eclipses of the Sun take place every 18 months somewhere on the Earth. After Freundlich's eclipse on August 21, 1914, there were total eclipses on February 3, 1916, and June 8, 1918, but no one was brave (or foolish) enough to attempt to collect data to support general relativity during the chaos of the First World War. By the arrival of the next total eclipse, though, on May 29, 1919, the scientific establishment believed that it was time to put Einstein to the test.[2]

The English astronomer and champion of general relativity Sir Arthur Eddington led an expedition to Principe Island, off the coast of Equatorial Guinea in West Africa.[3] It was a tense time for the observing team, as all eclipse work is. It only takes a few minutes of cloud cover to ruin the observation and make it necessary to wait many months for another good opportunity. Setting up their equipment in a mosquito-infested region, Eddington and his team waited nervously. The weather leading up to the eclipse was hardly encouraging. Day after day of thick cloud covered the island.

Things were no better on the eclipse day itself, Thursday, May 29. Still it was cloudy. The observers could only get their equipment ready to go and hope. Even as the Moon moved into place and began to eat away at the hazy disk of the Sun it was only possible to make out a faint shape through the cloud cover—there was no hope of detecting nearby stars. It was just minutes before totality that a gap opened in the clouds and Eddington and the team could get to work.

They managed to take sixteen photographs in the limited time available. As soon as the Moon shifted slightly from covering the Sun's disk, the stars were washed out by the emerging sunlight. But any relief that the team felt at capturing those images was shattered as the glass plates that held the images were developed. In shot after shot, the cloud that had hardly been visible to the naked eye was still present, still thick enough to conceal the faint stars that they were searching for. The plates were useless.

To make matters worse, toward the end of the run the clouds got thicker, removing any chance of making a measurement. In the end just two of the images were usable. But they were, according to Eddington, enough. The key stars had shifted out of position, by exactly the displacement predicted by Einstein and general relativity. It was a last-minute triumph. At least, that's the story as it is usually told. The reality was rather less satisfactory.

Even if general relativity had not been correct, the stars would be expected to shift. With his earlier special relativity Einstein had shown that energy and mass were equivalent. Although light has no mass, it has energy—and would be expected to be shifted by gravity even using standard Newtonian calculations. But the shift predicted by Newton was half that required for general relativity to be correct. Eddington had one plate that seemed to confirm Einstein, plus a second less clear image of which he said, "I think I have got a little confirmation from a second plate."[4]

In itself, a single reading with "a little" confirmation from a second was hardly definitive. Any single observation can be incorrect; it is certainly not enough to confirm a radical new theory. But in practice things were worse than having a single good result. Principe wasn't the only expedition to check the shift of the stars and test general relativity. There was a second expedition to Sobral in Brazil, which was also in the strip of the Earth that experienced the total eclipse.

The Sobral expedition had its own problems. Unlike Eddington, the observers in Brazil had clear skies, but the mirror on their telescope was distorted by the heat, producing fuzzy images. This made exact measurement impossible, but their best estimate was a figure much closer to that predicted by Newton than Einstein. They did have a second, smaller telescope that matched Einstein pretty closely, but the margin for error with this equipment was so large that its readings had no realistic value.

Taken as a whole, if all the data from the two expeditions were dispassionately analyzed, the only scientific conclusion was that it wasn't possible to either support or counter general relativity from the measurements taken. Later work demonstrated that this wasn't surprising. In 1962 an attempt was made with significantly better equipment to duplicate Eddington's findings during an eclipse—but even then it wasn't possible to be accurate enough to distinguish between Einstein and Newton.

However, Eddington was in charge in 1919, and he was totally convinced of the correctness of general relativity. He made sure that it was the Principe readings that were reported and dismissed those from Sobral. Arguably this was very poor science (though ignoring conflicting data is something many other great scientists, Newton included, have done), but as it happens Eddington's decision was correct. Since then there have been vast numbers of experiments that have confirmed the effects of general relativity in other ways, but at the time, the whole make-or-break decision resulting in headlines splashed across the world's newspapers was made more on intuition than honest interpretation of data.

Eddington's results were taken as confirmation of a measurement that was already known, but even easier to get wrong—the impact that relativity has on the orbit of the planet Mercury.[5] If you work out Mercury's expected orbit mathematically using the forces predicted by Newton's universal gravitation, the outcome doesn't quite match what is actually seen. The ellipse predicted by Newton shifts over time, or "precesses" in astronomical language. This shift is down to the influence of Jupiter (and to a much lesser extent other planets). Jupiter is sufficiently massive that it effectively causes tidal variations in Mercury's orbit.

Yet even with the most sophisticated correction for the influence on Mercury's orbit of all the planets, it still doesn't behave quite as it should. As early as the 1880s it became apparent that there was something wrong. The shift in the ellipse of Mercury's orbit was happening too quickly. It was even suggested there might be an unknown planet, lurking somewhere in space, perhaps permanently hidden behind another heavenly body, that caused this inaccuracy.

However, once Einstein had come up with general relativity, he was aware that the warping of space caused by the mass of the Sun would not just deflect light from distant stars, it would also influence the trajectory of the planets. This would be a relatively small effect, as the planets move much slower than light, but it should be noticeable. Specifically it should cause a shift in the orbital ellipses of the planets—exactly the "missing" amount from the variations in Mercury's orbit.

This was something Einstein knew about and could calculate

himself once his theory was in place. His equations corrected Newton and explained this deviation between theory and reality. After checking out the numbers he said, "For a few days I was beside myself with joyous excitement," and told a friend that he suffered from palpitations of the heart. But in itself the Mercury calculations weren't enough—he needed the confirmation of the eclipse observations.[6]

The drama of the story of the expedition to chase the eclipse added fuel to the publicity that surrounded Einstein and relativity. His public image was boosted to reach levels that would make him recognizable the world around for the rest of his life and through to the present day. Newton may be as famous a name as Einstein, but he can't match his twentieth-century rival for sheer recognizability. Like it or not, Einstein had become a true celebrity.

Just how much this was the case is clear in the way that Einstein was invited to appear for a season at the leading UK vaudeville theater, the London Palladium. Just imagine a present-day scientist being invited to take over some prime time TV entertainment show for a season merely to explain his theories. It was a remarkable offer. (Einstein politely declined.) But there was something extraordinary about relativity. It was surrounded by a feeling of mystery. It seemed somehow that this was the secret of the universe itself, something too complex for any lay person to understand, but that still sparked fascination and caused debate. In many respects Einstein and relativity had become an internationally acclaimed double act.

The equivalence principle makes it easy for us to expect light's straight line travel to be twisted into a curve, but such bending isn't the only effect that general relativity predicts that gravity will have on light. Imagine the light pouring out of a star like the Sun. As it escapes it is traveling under the influence of gravity, but it is moving directly away from the source of the gravitational field. What is that pull going to do to the light?

Let's bring the equivalence principle into play. We should see exactly the same thing happening to the light as we would see when accelerating. So first we need to work out what will happen to light when we move with respect to it. We know that light doesn't change its speed when we move toward it or away from it—that's the fundamental principle behind special relativity. But other things can change.

It's worth looking at light both as a wave and a stream of photons to get a clear picture. We know something very obvious happens to sound waves when we're moving. Everyone has heard the way a siren changes note as a police car heads toward us, passes us, and heads away. It is that same shift in frequency we hear as a train blasts past the warning bell of a railroad crossing.

When the car is heading in our direction, the sound wave it emits is squashed up—the waves come closer together because by the time a subsequent pulse is emitted, the car is that much closer. So the frequency of the wave is higher than it would be if the car wasn't moving. Frequency is just how often the waves arrive. When the car is heading away from us, the sound wave is stretched out—the frequency is lower than we would normally expect. So the siren note plummets as the car drives past. This Doppler effect can also be seen happening with light.

As we have already seen with moving galaxies, when a light source moves toward us, the frequency is pushed up—the light moves toward the blue end of the spectrum—it is blueshifted. Similarly a light source that is moving away shifts toward the red end. It's redshifted.

The picture is even simpler when we consider light as a stream of photons. When we are moving toward the light source, we give the photons extra energy (just as we get hurt more by a slow moving fist if we run toward it). When we move away from a light source, the photon energy is reduced. Higher energy photons produce a more blue light, and lower energy a more red light.

Now let's switch to a position above a star. The light is streaming toward us under the influence of gravity. When someone moves away from a light source, the light is red shifted, so by the principle of equivalence, the light should also be red shifted by the pull of gravity. If there were some way to switch off the gravitational pull of a star (and we could ignore all the other effects that this would have), its light would shift up the spectrum toward the blue.

One of the most striking implications of general relativity follows directly from this impact that gravity has on light, and to see it in action we need an imaginary device that was crucial to understanding special relativity's impact on time. This is a light clock. (Such clocks can be constructed, so they aren't truly imaginary, but we don't need a real light clock to undertake the thought experiment.)

Let's take a look at the special relativity version first. Imagine that we've got a pair of mirrors, exactly parallel, one above the other. We somehow set a beam of light going that hits the bottom mirror, bounces off at right angles, shoots up to the top mirror, bounces off, and so on, constantly flipping back and forth between the two mirrors. We can use this device as a clock. Each reflection by a mirror counts as a tick of the clock. Because light's speed is constant (assuming we don't change the material between the mirrors—let's make it a vacuum), the clock will never vary in its time keeping.

Now let's put this clock on a (transparent) spaceship that goes past us at high speed. As we watch we will see that the clock's light "pendulum" changes as a result of the motion of the ship. From our viewpoint outside the ship, while the light is heading from one mirror to the other, the ship will have moved at right angles to the light. This means that when it hits the bottom mirror, it won't hit at a point directly below the point it hit the top mirror. The line the light takes will be a diagonal.

This is a longer distance than the light is traveling from the viewpoint of someone on the ship. (For them it is still going vertically up and down.) The light's speed can't change, but for the outside observer it has traveled further, so its journey must take longer. The clock slows down from the outside observer's viewpoint. Clocks run slow if they are moving.

This is much easier to demonstrate with a vertical clock than it is with a horizontal clock where the beam of light flips back and forth in the direction of travel. There it would keep on the same path rather than take a diagonal, but because of the other effects of special relativity it would still slow down.

Now let's use that same light clock, at right angles to the movement of the ship, to take a look at the effects of general relativity. An observer onboard the constantly moving spaceship used for special relativity would not see any deviation of the light away from the vertical. But if the ship is accelerating, the mirrors will move sideways with respect to the light and the onboard observer will see the light beam deviate sideways. The path will be extended, so the clock will tick slower. Clocks slow down under the influence of acceleration and so the equivalence principle tells us that they will also slow down under the influence of gravity.

Another way of looking at this that makes more direct use of the red/blue shift of light by gravity is to think of a beam of light being shot downward toward the Earth. Equivalence tells us the light should increase in energy. It can't go up in speed, because it always goes at the same speed, but remember the energy of a photon is proportional to its frequency. It will be blueshifted as it "falls," increasing its frequency. (Just as it would be redshifted climbing away from the planet.) This means a clock timing the frequency of the light up top will have more ticks per wave of the light than an identical clock timing the frequency down at the bottom. The clock experiencing a higher gravitational pull will run slower.

These two relativistic effects on time are not just nice theories; they have profound influences on cosmology, and even have practical implications. GPS satellites, for example, rely on comparing the time signals from a number of satellites to establish location. But the clocks on those satellites generating the signals are subject to the influences of relativity. Firstly the satellites are moving quickly—around 87,000 miles (14,000 kilometers) per hour. So special relativity effects mean the clocks run slow, by about 7 millionths of a second per day.[7]

Secondly the GPS satellites orbit at around 12,400 miles (20,000 kilometers) up. At this height they experience a lower gravitational pull than we do on the surface. That means the clocks on the satellites run faster than they would on the Earth, where the higher gravitation would slow them down. Because of this, the clocks run fast by about 45 millionths of a second per day. Overall time on the satellites gets ahead, gaining around 38 millionths of a seconds per day. This sounds trivial, but if not corrected, the GPS system would be out by around 6 miles (10 kilometers) in just a day.

So far we have seen general relativity as the idea that mass warps space-time and so produces effects that are the equivalent of a force, though as far as the objects being influenced by gravity are concerned, they do not experience any force, just travel unaffected in a straight line and it is really space-time that is changing around them. This is an awesome conceptual leap of understanding that Einstein produced, but it isn't what took him 8 years of work. The real effort went into producing the math to support the picture.

It's certainly true that this picture of Einstein battling for 8 years against the difficulties of taming the mathematics needed to understand gravitation is a popular one, but as Einstein's biographer and friend Abraham Pais points out, it is not entirely borne out by the evidence. Einstein came up with his "happiest thought of his life" in 1907 while still working at the patent office in Bern. He took on some lecturing duties soon after, but in 1909 he started work in his first proper academic role as an associate professor at the University of Zurich, and then in 1911 became a full professor at German Karl-Ferdinand University in Prague.[8]

Without doubt, part of this early period of the gestation of general relativity was spent on the everyday time-consuming effort of moving house and on his teaching work. Einstein also had a growing family, with a second son Eduard born in July 1910. (Sadly, Eduard was to suffer from schizophrenia and died in an institution in 1965.) But most important, he had other aspects of science on his mind. The equivalence principle, linking acceleration and gravity, was still a fascinating observation but not one that immediately drove him into serious work.

Instead, it seems likely that most of Einstein's original thinking between 1907 and 1911 was focused on quantum theory. His 1905 paper on the photoelectric effect (the one that would win him the Nobel Prize) had stimulated much thought and debate on quantum physics. It changed what eventually would be called the photon from being a convenient mathematical trick to a real particle. And this was beginning to raise all the strange probability-based implications of quantum theory that Einstein came to hate.

While it is not possible to be absolutely certain what Einstein spent his time thinking about, all the evidence points to quantum theory, both in terms of his publications, but also, most significantly, in his letters. Einstein was never one to keep his scientific problems to himself. He liked to discuss the challenges that faced him in writing with friends. And throughout this period it was quantum theory that seemed to be nagging away at him.

In June 1911, though, he notes returning to gravitation and by 1913 had produced a paper that was a halfway stage to general relativity. It's possible that the isolation of Prague helped in this. In some ways it was a strange move. Zurich had been a very comfortable location for him—after all, he had chosen Swiss nationality and made his home in Switzerland when all of Europe lay before him. The Swiss authorities had been increasingly aware of what they might lose, and had made him a number of offers of promotion before he left for Prague.

Although his new Czech university was German speaking, he seems to have had few colleagues to confide in there. Prague wasn't much of a center for physics, and he found the rather stiff, formal culture alien to his relaxed nature. Perhaps he had been too comfortable in Zurich to really concentrate on new thinking, and the lack of other distractions in Prague opened up the possibility of taking on a challenge of the scale of general relativity.

If the move to Prague got him thinking about gravitation, it was not where general relativity was truly born. In less than 2 years Einstein was back in Switzerland, tempted by an offer from an old friend, Marcel Grossman, the man whose lecture notes had helped Einstein through his degree. Grossman was now the dean of the mathematics and physics faculty of Einstein's alma mater, the ETH, and it was to teach there that Einstein returned in 1912.

It seems this move was perfectly timed. Einstein was struggling with finding an appropriate mathematical tool to deal with the complexities of the implications of space-time warped by mass. Yet within 6 days of arriving back in Zurich he was to write, "It is going splendidly with gravitation. If it is not all deception, then I have found the most general equations."[9]

The realization that had struck Einstein at this stage was the need to understand non-Euclidian geometry, the geometry of space that isn't flat. The basic geometry we learn at school always takes place on flat surfaces. We are told, for example, that the angles of a triangle add up to 180 degrees, and parallel lines never meet. Yet all this goes out of the window on a curved surface. Drawn on a sphere, as we have already seen, parallel lines do meet, and the angles of a triangle exceed 180 degrees.

This, Einstein knew perfectly well, and he had investigated some non-Euclidean approaches, including the theory of curved surfaces developed by the German mathematician Carl Friedrich Gauss in the 1820s. But Grossman put Einstein onto a particular geometry of curved space known as Riemannian geometry after the German mathematician Bernhard Riemann. This would give Einstein the key tool that he needed to explore gravitation, and a joint paper with Grossman displays many of the concepts that would eventually make it into general relativity.

Now, yet again, his personal and academic circumstances would change. The ETH would have been a natural home for Einstein, were it not for the teaching requirement. But Einstein had been approached by the great German scientist Max Planck to see if he could interest him in a post at the University of Berlin, where he would have the great benefit of not being obliged to teach. Planck was the grand old man of German physics, and Einstein was no doubt flattered by the support he was given by such a grandee. Coupled with a strong financial package and personal reasons that would only be revealed later, this was enough to put the Einstein family on the move yet again.

This move seems to have been the last straw in what was already a troubled relationship. Although his wife Mileva and their children joined him in Berlin it was only a few weeks before they were split up, his family returning to the more familiar Zurich, leaving Einstein briefly on his own. A contributory factor might well have been Einstein's closeness to his cousin Elsa Lowenthal, who would become his second wife. He would later refer to his cousin as the person who "in fact drew me to Berlin."[10]

After more stumbling over the math and hitting several dead ends, it was in Berlin on November 25, 1915, that Einstein presented a paper to the Prussian Academy of Sciences with the title "The Field Equations of Gravitation." General relativity had reached its final form. Einstein's masterpiece, the paper that would push even his great works of 1905 into the shade, was in print.

Einstein had produced a series of equations known as the Einstein field equations that describe mathematically how general relativity will act. The math involved is fiendishly complex—too complex for Einstein initially, which was why he had to get help to understand the tools he would need to produce these equations. What's more, just having the equations is not in itself the end the matter. Those equations need to be solved.

Solving an equation is something that we are introduced to in high school mathematics. We might have an equation like $x^2+2x-8=0$. Knowing that this is the appropriate equation for a particular situation is one thing—knowing what the solutions to the equation are is another. You may remember that for this particular type of equation, the quadratic, there will be two values of x. We can transform the equation into the form $(x+4)\times(x-2)=0$. We then turn either of the brackets into 0, making x either −4 or 2. These are the solutions.

The solutions to Einstein's field equations for general relativity are vastly more complex than a quadratic equation, and rather than a complete solution, they have only proved soluble for specific cases which simplify the equations to a more manageable form.[11] (This would be like, for example, solving the quadratic equation only for cases when x is greater than 0.) Yet despite this complexity, amazingly the equations of general relativity can be condensed to a single line of text:

$$G_{\mu\nu}+\Lambda g_{\mu\nu}=(8\pi G/c^4)\ T_{\mu\nu}$$

It's quite remarkable that such a powerful piece of science—an explanation that so far copes with everything that gravity can throw at us, that predicts all the subtleties of warped space and time—can be condensed into such a compact form. Mathematicians and physicists often refer to the "elegance" in an equation or theory—even without a clue to what the details of the equation mean. There is surely something very elegant about Einstein's work. It's a formula you could (and do) find on a T-shirt.

At first glance it's easy to wonder what all the fuss is about. This is the equation Einstein had to get help with the math on? It looks trivial. The part that looks most complex is actually just a constant. That bit in brackets ($8\pi G/c^4$) involves three well-known constants—pi, the universal gravitational constant G, and the speed of light c. Yet despite appearances, this is by far the simplest part of what you can see.

The reason for this is that apart from another constant Λ (of which more on page 278), each of the other three components of the equation aren't what they seem. They are mathematical wolves in sheep's clothing. They are neither constants nor variables like the x in our quadratic equation above. They are tensors—and it was dealing with these that forced Einstein to get help to be able to formulate the mathematics of general relativity.

A tensor is a mathematical object, potentially in many different dimensions, that can be anything from a multidimensional matrix of values, to a series of complex equations in its own right. Einstein's tensors collapse ten complex equations into a single entity and it is the complexity lying behind the tensor representation that causes problems for anyone attempting to solve Einstein's equations. To make matters more difficult, these are differential equations, equations that relate to values that change with time and place, not just simple fixed numbers.

The problem with this notation is that unless you understand what lies behind the symbols, they have no direct meaning. To give a simpler example to show how this kind of mathematics works, imagine that we have some kind of force which depends on the square of the distance plus a constant. We would write this conventionally as

$$F=r^2+a$$

where r is the distance and a is the constant. But we could also write it as

$$F=f(r)$$

where f is shorthand for “multiply the thing in brackets by itself and add a.” Looking at f there is nothing to tell us what the equation lying behind it is. It is simply a shorthand to avoid having to write out the equation over and over again. In this case f is called a function, something that would be familiar to any computer programmer.

In the case of Einstein’s tensors, things are more complex, with those three symbols hiding ten complex equations, but the effect is the same.

Although the derivation of those equations is too messy to step through, what we can do is see the influences that made those tensors necessary. For Newton there was just one influence on gravity—mass—but Einstein was bringing relativity into the picture, and with relativity things get more complicated. In total, Einstein would discover four different sources for a gravitational effect to incorporate into his equations.

The first component is the mass of the body producing the space-time warp—but it's a case of mass plus the other contributors that relativity brings into the equation. The mass contribution is not just the mass we would measure if an object was stationary. It also includes any energy that object has, for example due to motion, because mass and energy are interchangeable under relativity. More surprisingly, it also includes pressure.

Pressure is just force applied to an area. In a gas, the pressure comes from the many molecules of the gas repeatedly battering the walls of the container (or whatever the gas is providing pressure on). If you look at something like the Sun there is plenty of pressure in the star's interior from the energetic movement of the particles inside. This pressure also contributes to the basic "mass" gravitational effect. In many normal circumstances, the pressure component is negligible, as it turns out to be divided by the square of the speed of light in its impact on gravity. But once we get onto stars and other special bodies, pressure becomes significant.

This "basic" mass-derived effect of gravity results in a curve in time. This is needed for the familiar aspects of attraction. As we have seen, when time is warped, a change in time becomes a change in spatial position—over time something will begin to move. The warping of time by mass and pressure produces the basic acceleration under gravity.

But we also see a second effect that was detected in the Principe solar eclipse. Here it is space that is being warped to transform the straight beam of light into a curve, and the aspect of matter that produces this warp is a lot more complex than the basic combination of mass and pressure. There we were dealing with warping of a single dimension, time, but here there are combinations of three dimensions, each with two potential directions, providing six different factors contributing to the spatial curvature. Each of these factors is still influenced by mass, energy, and pressure, but the combination makes for a more complex mix.

The third component of general relativity is one that emerges from special relativity. The need for it can be seen from a simple example, though the implications are quite messy. Just imagine we have an object sitting between two very long, identically heavy strips of material out in space. The top strip is moving from left to right and the bottom strip from right to left. Each strip travels at the same speed.

We wouldn't expect the object sitting in the middle to move toward either strip because they have the same mass, so the gravitational pull on the object will balance out. But now let's fly past from left to right in a spaceship, going at the same speed as the strips. (Relativity wouldn't be the same without spaceships.) From our viewpoint, the "stationary" object is now moving from right to left. The top strip isn't moving at all. And the bottom strip is moving at twice the speed it was before.

So from the spaceship, the bottom strip is moving but the top isn't. Special relativity tells us that a moving object has increased mass. So from our viewpoint the bottom strip has more mass than the top strip. This means that the object should be pulled toward the bottom strip, because there will be a bigger gravitational force as a result. Yet in practice the object doesn't move up or down. Our traveling past it sideways won't make it start moving up or down.

There is only one conclusion that can be drawn. Remember that from the spacecraft, we see the object moving from right to left. That is the only other thing that has changed in the setup. So the fact that the object is moving right to left must cause a gravitational pull at right angles to that motion that cancels out the pull of the heavier moving strip. The object is moving from right to left, causing a gravitational pull at 90 degrees to motion, in this case upward.

This is sometimes referred to as gravitomagnetism. It's not that gravity is generating magnetism—the effect has nothing to do with magnetism. It's just that this gravitational force is acting on an object in a similar way that magnetism is generated by a moving object with an electrical charge. But we are talking about a purely gravitational effect, the third component that Einstein had to incorporate, which means that moving objects feel a (small) sideways displacing gravitational pull.

When this effect occurs with a rotating object, like a planet or star, the process is known as frame dragging. It can be seen as a bit like rotating a spoon in a jar of honey. As the spoon rotates, it pulls the nearby honey with it, dragging the gooey substance into a vortex. Similarly, a rotating body drags space-time with it because of this sideways gravitomagnetic effect.

This frame-dragging effect has the potential to produce some surprising outcomes. If a very massive object is rotated at high speed, space-time should be dragged with it to such an extent that it effectively forms a time machine, making it possible to take a journey through space that results in traveling through time.[12] Many of the theoretical constructions of time machines that allow travel into the past are based on frame dragging and the gravitomagnetic effect.

Frame dragging was for many years a result that was predicted by general relativity, but had not been demonstrated, as the effect is relatively weak. One confirmation of it finally arrived when NASA's longest running project delivered its preliminary results on the gravitomagnetic effect in 2011.

The project was Gravity Probe B, an experiment that was planned in 1962 and has run for 49 years, led throughout by Stanford physicist Francis Everitt, truly a life work. The idea was to put a gyroscope into space and use it to look out for the gravitomagnetic effect, which ought to push the gyroscope sideways over time as it spins. The satellite was originally launched into a polar orbit in 2004, with a total of four gyroscopes on board.

These were not the traditional flywheels on gimbals, but rather four quartz spheres coated in superconducting niobium, the most perfectly spherical manmade objects ever created according to *The Guinness Book of Records,* which rotated at around 5,000 revolutions a minute.[13] The idea was to use the magnetic field in the superconducting surface to measure the precise orientation of the gyros. Unfortunately tiny imperfections on the surface of the spheres (and even more so on their mounts) caused unexpected behavior from the gyroscopes, which didn't act in the smooth way expected.

By 2007, the Gravity Probe B team had managed to demonstrate the basic warping of space-time by gravity, but not the much weaker effect of frame dragging. The team hoped to remove the influence of the imperfections from the data to uncover the effect beneath, but this involved a vast amount of work and in 2008 NASA decided to cut its losses and stopped financing the project.

Luckily for those who had stayed with the Gravity Probe B for so many years, private funding enabled them to continue working, undertaking an amazing piece of remote forensics in working out just what the corruption to the data was, and how to remove it. By 2011 they were ready to publish confirming results.[14]

By then, the team was no longer the first. Another project had used the frame dragging experienced in the orbits of two satellites that were tracked with laser rangefinders to determine how much influence general relativity's gravitomagnetic effect would have on them.[15] These results came out in 2004, confirming Einstein's predictions. And more accurate measurements are expected from the Italian Laser Relativity Satellite, due to be launched in 2011. Yet there is something admirable in the doggedness of the Gravity Probe B team to achieve their results—and it provides excellent confirmation.

Frame dragging was not the last of Einstein's components in assembling general relativity. Finally, and perhaps most strangely, Einstein had to deal with a gravitational positive feedback loop. We're all familiar with positive feedback when a microphone gets too close to a speaker, producing an ear-splitting howl of noise. This is runaway positive feedback—but not all positive feedback is like this. The distinction between the two types is like the difference between these two series of numbers:

$$1+\frac{1}{2}+\frac{1}{3}+\frac{1}{4}+\frac{1}{5}+\frac{1}{6}\ldots$$

and

$$1+\frac{1}{2}+\frac{1}{4}+\frac{1}{8}+\frac{1}{16}\ldots$$

They look very similar and each could describe the way the feedback is working, where each fraction is the level at which feedback occurs. However there is a huge difference between the two series of numbers. The first one is runaway. If you were to add up all the fractions in the series you would get an infinite total. In practice, of course, with the microphone and speakers, the output is limited by the capacity of the amplifier, but in principle it would build up to unlimited power.

The second series is more subtle. Here, even if you add in a whole infinite set of fractions, you end up with a finite number. The total is just 2. And this is more like the feedback loop that Einstein had to deal with in the fourth contribution to general relativity. This arises from the counterintuitive proposition that gravity itself has a gravitational pull.

Think back for a moment to high school science. We were taught about two important kinds of energy—kinetic energy and potential energy. (Yes, there was heat and chemical energy and nuclear energy as well, but let's just focus on these two.) If you imagine pushing a large boulder to the top of a hill, it takes work, which is just energy being transferred from place to place. You put the work into the system and the result is that the boulder gains potential energy. (A lot of your work will probably go to friction, but let's not worry about that.)

You then let the boulder go. It rolls down the hill. The potential energy is converted into kinetic energy, the energy of motion down the hill. That kinetic energy had to come from somewhere—the potential energy was real, even though it's something less obvious than kinetic energy. And what causes potential energy? Gravity. Without gravitation, the boulder would just sit there. It would feel no urge to move down the hill.

So gravity is a source of energy. But in Einstein's relativistic world, energy and mass are interchangeable. Any amount of energy is the equivalent of an amount of mass. So the energy present in a gravitational field will itself generate a small additional amount of gravity, as a result of its "mass." And this extra gravity will itself generate an even smaller amount of energy and so on. This isn't a runaway feedback loop, so the result is a small increment to the expected gravitational force that would mean the final value was underpredicted unless this was taken into account.

Before we leave the basics of Einstein's work behind and explore the more exotic implications of general relativity, let's take the equivalence principle on a final outing to see if you've really got the hang of it. Imagine you are driving along at a good speed in a car. You are sitting in the passenger seat, holding a helium balloon by its string, which stops the balloon rising to touch the ceiling of the car. Suddenly, the driver brakes heavily. What happens to the balloon?

Without using the equivalence principle, common sense will probably tell you what is going to happen. The car decelerates rapidly as the brakes are applied. There is no similar deceleration applied to the balloon, so from the balloon's viewpoint, the car moves vigorously backward. The result is that the balloon should head toward the front of the car. Unfortunately, as is often the case with physics, common sense is wrong and the equivalence principle explains why this is the case.

The equivalence principle tells us that acceleration has exactly the same effect as a gravitational pull in the opposite direction. So, for example, in the spaceship we've used frequently up to now, an upward acceleration of the spaceship is felt as (and is the equivalent of) a downward pull of gravity. In the balloon example, the car is accelerating backward, so let's use equivalence to convert that into a gravitational effect.[16]

The idea that the car is accelerating backward sounds a little odd. We normally think of the car as going forward, but decelerating. However a "deceleration" is just an acceleration in the opposite direction of travel. Acceleration is a change of velocity. In this case the change of velocity is in the opposite direction to the way the car is traveling. So slamming the brakes on produces a backward acceleration. And according to the equivalence principle this is just the same as feeling a gravitational pull toward the front of the car.

Of itself this is no surprise. We know that when the driver slams on the brakes we are thrown forward against our seatbelts as if we are feeling a pull of gravity toward the front of the car. Now here's the thing. What does a helium balloon do as a matter of course? It floats in the opposite direction to the pull of gravity. That's the whole point of a helium balloon. Its buoyancy means that it moves in the opposite direction to the gravitational force. The helium is lighter than air, so the balloon feels an upthrust against gravity.

So now we've got the full picture we need to understand what will happen to that balloon. Briefly it may appear as if the balloon is moving toward the windshield because your backward acceleration will mean you pull your end of the string toward the back of the car. Imagine holding a balloon and suddenly pulling on the string horizontally. From your hand's viewpoint the balloon moves in the opposite direction.

After this immediate effect, though, the balloon will float off against the pull of gravity. And in this case the "gravitational pull" is due to the acceleration and the equivalence principle. The gravitational pull is toward the front of the car, so the balloon will head toward the back of the car—in the totally opposite direction to the one predicted by common sense.

With a good grasp of general relativity—a theory that would turn out to have many more implications, including providing our best understanding of the way the universe began and changes over time—we can fit gravity into a bigger picture of the physical world. Gravity is a fundamental force, but it's not the only one. It has three cousins to contend with.

CHAPTER EIGHT

ONE OF FOUR

> *It will be found that everything depends on the composition of the forces with which the particles of matter act upon one another; and from these forces, as a matter of fact, all phenomena of nature take their origin.*
>
> —*Philosophiae Naturalis Theoria* (1758)
>
> Roger Joseph Boscovich

Gravity, as the medieval philosophers were aware, is not alone as a force in the universe. They also knew about the magnetic and electrical effects produced by the electromagnetic force. For a long time there was a suspicion that gravity and magnetism were just variants on a theme. Admittedly, magnetism has some significant differences. Not all objects are influenced by magnetism, while all are by gravity. Gravity always appears to be attractive, while magnetism can be attractive or repulsive, but both appeared to involve a mysterious interaction at a distance.

Early attempts to explain the nature of gravity often put it down to some special kind of magnetism. The possibility of a connection was still in the air in the nineteenth century when the great English physicist Michael Faraday was at work at the Royal Institution in London. In 1849, Faraday wrote in his notebook: "Gravity. Surely this force must be capable of an experimental relation to electricity, magnetism, and other forces . . ."

Faraday went on to undertake experiment after experiment to try to discover some kind of relationship between the two forces. He knew, as he had proved, it was possible to make magnetism from electricity and electricity from magnetism. Why not gravity from electricity or magnetism from gravity? But unlike his many successes with electromagnetism, he could not get gravity to play ball. In the end he had to conclude, "Here end my trials at present. The results are negative. They do not shake my strong feeling of the existence of a relationship between gravity and electricity, though they give no proof that such a relation exists."[1]

As we will discover, gravity and electromagnetism proved not to be the only forces responsible for actions in the universe, but the urge to find some linkage between these fundamentals that shape reality has not gone away. After his huge contributions to science in 1905 and the development of general relativity 10 years later, Einstein would spend many years looking for a mechanism to produce a "unified field theory" (we'll come back to why it was a "unified field theory" rather than a "unified force theory")—a search that proved fruitless. He would produce no more significant results after general relativity.

Faraday's experimental genius and Einstein's unparalleled grasp of theory both failed to find a link between gravity and the other forces. It seems that gravity stands alone. It might seem obvious, in that gravity has a very different mechanism if it is truly about warping space-time. In effect, gravity isn't a force at all, but a side effect of this distortion. But Einstein hoped that he could find a similar warp-based explanation of electromagnetism, turning it, too, into an effect of space-time geometry. As yet there is no evidence that this is possible.

The full set of forces are gravity, electromagnetism, and the strong and weak nuclear forces. We have already met the first two. As for the others, the strong nuclear force is responsible for binding together the quarks that make up particles like the proton and the neutron, while a leakage of this force binds the nucleus together, overcoming the electromagnetic repulsion of the positively charged protons. The strong force only acts over short ranges, limiting the possible size of an atomic nucleus.

The weak nuclear force comes into its own during nuclear reactions. Even shorter range than the strong force, this weak interaction requires particles to be a tiny fraction of the diameter of a proton away from each other. The weak force acts as a switch for quarks, changing them from one flavor to another. The result is that nuclear particles can change type, as when a proton switches into a neutron in the nuclear fusion reactions in a star, or during nuclear decay processes, like beta decay, producing high-energy electrons from the nucleus.

These famous four forces are often defined in terms of fields. These fields are somewhat like the contours of a piece of land—they are continuously varying properties (corresponding to the height of any point on that land) that fill space. Perhaps the most familiar field was Faraday's discovery, the magnetic field, easily demonstrated by sprinkling iron filings on a sheet of paper over a magnet, producing lines running from pole to pole of the magnet that map the way the field progresses.

It's tempting to think of fields as useful mathematical concepts that make it easier to do many of the calculations that are required for physics, but that don't have any reality. They seem more abstract, less tangible than the idea of a force being carried by a particle. It seems more straightforward to imagine the electromagnetic force being carried by photons than to consider it a field that stretches out indefinitely into space.

This seems to be because the idea of a particle carrying a force has direct parallels in the "normal" world we experience. If I want to give someone at a distance away a push, I can throw a softball at him or her, and they feel the force, literally. Fields don't have that same level of parallel in the world we experience.

However, tempting though it is, to think of fields as imaginary mathematical constructs while photons, say, are concrete objects, is misleading. A photon is no more tangible than a field. I can't show you a photon or a field. I can detect either a photon or a field using the right equipment. To say that one exists and the other doesn't is to apply the worldview of the macro world to the submicroscopic. Thinking of particles is the only sensible approach when dealing with an individual photon, but when dealing with vast quantities of photons, a field is often easier to work with.

The closest we have to a real world parallel for a field is an undulating landscape. Just think of an area of land featuring a steep cliff. You stand at the bottom of that steep cliff. I have a softball in my hand. I can drop it on your head as I stand next to you, or climb to the top of the cliff and then drop it on your head. Which hurts more? Clearly the ball dropped from the top of the cliff.

But why does it hurt more? Your natural feel for what is happening here more naturally parallels the concept of fields. You don't say that there are more opportunities for force carrying particles to work on the ball dropped from the cliff, you say it had more potential energy when it was up on top. As it fell, that potential energy was converted to kinetic energy. This potential energy is measured by the contours of the ground. Higher points will produce greater energy. It's an argument that has the feel of a field.

It should be possible to explain gravity in terms of carrier particles called gravitons—in fact it is going to be necessary if we are to develop a quantum theory of gravity (see the next chapter), but our most accessible understanding of it comes from the "classical" (non-quantum) general relativity. As we have seen, that is all about warping space-time. With such a picture, fields are a more accessible approach than carrier particles.

Gravity is weak—and yet it is a force that can achieve some remarkable natural feats. Take, for instance, the neutron star. Neutron stars are collapsed stars composed entirely of neutrons. Usually the nuclei of atoms are a mix of protons and neutrons, but under the immense pressure of the collapse of a star, protons and electrons combine to form extra neutrons.

In effect, a neutron star is a single atom of immense proportions, but a very strange atom. Standard atoms are limited in size by the range of the strong nuclear force. The biggest naturally occurring atom is uranium with 92 protons and between 141 and 146 neutrons in the nucleus. Artificial atoms have been constructed with larger numbers of protons and neutrons (collectively known as nucleons) but they are all unstable.

By comparison with the strong nuclear force, electromagnetism has a much greater reach. So as more and more protons are piled into the nucleus you have growing electromagnetic repulsion between the positively charged protons. As the nucleus gets bigger and bigger, the strong nuclear force runs out of steam. It can't reach from one side of the nucleus to the other. Result? The nucleus blows apart.

The neutrons in a nucleus don't have the same problem. As far as electromagnetism goes, you can add them indefinitely. There is no repulsion because they have no charge. But there is another problem here—the Pauli exclusion principle.

This is an aspect of quantum theory that becomes very important when some types of particle get together. Quantum particles are divided into two types: fermions and bosons. Technically the distinction between fermions and bosons is their spin. This is quantum spin—not a measure of actually spinning around, but a quantum property that all particles have and that comes in units that are rather irritatingly halves.

Bosons have spins with whole integer values (0, 1, 2) while fermions have half integer spins (½, ³⁄₂, ⁵⁄₂). But what is relevant here is that bosons and fermions have different behaviors when it comes to working together. Bosons are the herd animals of the particle world, while fermions are loners. A collection of bosons can all be in exactly the same quantum state, but fermions have to be unique.

This "quantum state" is the collection of properties of a particle—if two fermions are in the same place (and we have to be a little careful about what we mean by "same place"), then they have to have different values of at least one property. When I was first introduced to the idea of the Pauli exclusion principle at university I was puzzled by this idea of the particles being in one place. Surely there wasn't a suggestion that they were all sitting inside each other like a Russian doll?

What we have to bear in mind when thinking of particles occupying the same space is that quantum particles are generally in more than one place at a time. Unless we make a measurement, they exist in a spread out form, with different probabilities of being in different places. The greater the overlap of these probabilities, the more particles can be said to be in the same place.

Among the fermions we, find all the familiar components of matter—electrons, protons, and neutrons. (Bosons are the likes of photons and the gluons that hold quarks together.) So finally we can see why we can't just pile in the neutrons indefinitely. Once you have a few neutrons occupying the lower energy states, you have to start adding neutrons with more and more energy. If they are close enough together, the Pauli exclusion principle insists on this. And with more energy they have a natural tendency to fly apart.

The only hope of keeping large numbers of neutrons together is to have strong enough gravity to hold them in. That means accumulating a fair amount of mass. So once you've gone beyond a standard atom size, you need around 10 percent of the mass of the Sun—perhaps 2×10^{29} kilograms (4.4×10^{29} pounds)—before gravity is likely to be strong enough to hold a collection of neutrons together. There is also an upper limit of around twice the mass of the Sun, above which gravity overcomes the pressure in the nucleus and further collapse follows, forming a black hole.

This doesn't mean you can have neutron stars as small as 0.1 times the mass of the Sun, simply that if you could get that mass of neutrons together, they would be stable. But you have to accumulate those neutrons. A neutron star forms from a collapsing star. Up to a certain mass, called the Chandrasekhar limit (around 1.4 times the mass of the Sun), there is too much pressure arising from the different energies of the fermions in the star. Above that mass, gravity can take over and the star will collapse into a neutron star.

This delicate balance is one that is hugely important as far as human existence is concerned. You will often hear about the "fine tuning" of the universe, the idea that certain parameters of the physical universe had to be just right (what is sometimes described as a Goldilocks universe) before life as we know it could exist. But the importance of neutron stars for life is not something that is often brought up.

Neutron stars are amazing, certainly. Ordinary atoms are mostly empty space. The nucleus in the middle of the atom has been likened to a fly in a cathedral, with the rest of the space empty but for an insubstantial cloud of electrons. But neutron stars throw away all that empty space—they are pure nucleus. In a single cubic centimeter—little more than the size of a sugar cube—a neutron star contains around 100 million tons of matter. An entire star, heavier than our Sun, occupies a sphere that is roughly the size of Manhattan.

However neutron stars aren't just curiosities in the stellar zoo, they are essential parts of the production line of heavy atoms. Just imagine that neutron stars didn't exist and that instead, all collapsing stars became black holes. Neutron stars are usually formed in a supernova. As the neutron star collapses, the outer layers are blown off in an immense explosion. All this matter goes flying off into space. And such supernovas are the source of all the heavier elements on the Earth. We simply wouldn't be here were it not for supernovas like this.

When a black hole forms there is no such explosion. Everything delicately slides into the inescapable gravity pit. It is only thanks to neutron stars and their explosive formation that we exist.[2]

I've mentioned black holes many times already without bothering to explain what they are. This is because they have become part of the language, an accepted norm, a handy way of referring to a bottomless pit that swallows everything up and never lets anything out. It's arguable that black holes have such a strong grip on the imagination because they have become mythic. These are mythological objects with remarkable powers—the dark, all-consuming spirits of space.

It's impossible to start with a clean sheet when thinking about black holes as there are just too many preconceptions. Some are accurate—that nothing can get out of a black hole once it gets past a certain limit, for example. Some are almost right—that a black hole is black, springs to mind. In practice they can produce vast quantities of radiation directly and indirectly, but it is true that an isolated black hole with nothing else involved would be perfectly black. And then there are the ideas that are downright wrong.

Hollywood tends to portray black holes as both irresistible forces that nothing can approach and as gateways to other universes. Although each of these myths is based on a truth, each is way off the mark. If you were in a spacecraft, happily orbiting around a star, and the star suddenly turned into a black hole, Hollywood would have you panicking as a whirlpool of gravitation suddenly began to pull you inexorably toward your doom.

In reality, you would carry on without noticing, calmly orbiting just as before. (At least as far as the gravitational pull is concerned. A black hole's formation could produce some nasty radiation.) A black hole is a collapsed star—but it has no more mass than the original star. Its gravitational pull at a safe orbital distance will be exactly the same as was the case with the original star. What is different, though, is that you can get much closer than you otherwise could.

With a regular star you can't get too near because of the intense radiation and heat—a bombardment that would keep you significantly farther from the surface of the star. This surface would be at least 1.5 million kilometers (930,000 miles) from the star's center in the parent star for a black hole. Once the star forms a black hole, you can approach as close as you like, experiencing a stronger and stronger gravitational pull. The radius of the event horizon, the point of no return, for a typical stellar black hole would be just 15 kilometers (9.3 miles).

We will look later at the ways that make it possible to use a black hole as a portal, but for a regular black hole with no extra features, this just isn't possible. A black hole is a one-way street with no exit. If you could survive, yes, you could in principle travel an infinite distance within the black hole—but you would never come out again.

In practice, approaching most black holes would be deadly. Firstly, you would be bombarded with extremely powerful radiation. Any particles that stray too near the black hole will be accelerated toward it, generating high-energy radiation if those particles are charged. You are likely to be hit both by the radiation and any passing particles.

To make matters worse, the closer you get to the black hole, the more the tidal forces are going to give you problems. Imagine you are heading toward the black hole feet first. Your feet, being closer to the center of gravity, will feel a stronger gravitational pull than your head. This is the same effect that makes it possible to get around the principle of equivalence if you are inside a spaceship. But in the spaceship example, the difference in gravitational pull will usually be quite small.

Approaching a black hole, it's different. Get close enough and there will be hugely more pull on your feet than on your head. So just like the falling stones (see page 109), pulled apart by the stronger gravitational pull nearer the massive object, your feet will be pulled away from your head. Inexorably you will be stretched by the ultimate instrument of torture. In a process given the delightful name of spaghettification, you will be turned into a long, thin, strip of matter.

But we are getting ahead of ourselves. Black holes have had two birthplaces, once in the eighteenth century and again in the twentieth. In neither case was this a matter of astronomy and observation. The black hole is first and foremost a construct that emerges from theory—but we have no reason for thinking the theory to be wrong, and there have been observations since the second origin of the concept that seem to support black holes' existence.

The eighteenth-century black hole was a product of Newtonian gravitational theory. English astronomer John Michell, born in 1724, was considering the idea of escape velocity.[3] This is a straightforward consequence of Newton's work. If I throw a ball into the air, as soon as it leaves my hand the only force acting on it (apart from air resistance) is gravity. From the moment it leaves my grip it is being accelerated toward the ground. The result is that its climb gets slower and slower, eventually stopping and reversing to fall back to Earth.

However, we know from Newton's law of universal gravitation that the strength of the force of gravity drops with the square of the distance away from the center of the Earth (or whatever is causing the gravitational pull). If I can get a ball 3,958 miles (6,370 kilometers) above the surface of the Earth, I have doubled the distance from the Earth's center, so the pull of gravity will be down to a quarter of its usual value.

If I can throw a ball fast enough, the pull of gravity will drop off faster than the ball decelerates. The speed at which I need to throw the ball to avoid it ever dropping back to Earth is the escape velocity—from the surface of the Earth this is around 11.2 kilometers per second (25,000 miles per hour).

Escape velocity is very easy to misunderstand. Here's a statement from an otherwise excellent textbook on gravity: "Getting away from Earth to another planet therefore must require a launch speed with greater than 11.2 [kilometers per second]."[4] This is just not true. You only have to watch a rocket taking off from Cape Canaveral to see this. They do not launch at 11.2 kilometers per second. If they did, they would be out of sight in a second. Instead they claw their way slowly into the sky.

A spacecraft can leave the Earth at walking pace if desired. This is because it is under power during its flight. Escape velocity assumes that the only force applied to the object after it leaves the surface is gravity. In reality, a space rocket has a huge motor applying an upward force, and so the 11.2 kilometers per second figure isn't relevant. This is only a limit for a projectile that is not under power.

Back in the eighteenth century, Michell was calculating the escape velocity of different bodies. At the time he was working, it was already known that light had a specific speed, around 300,000 kilometers (186,000 miles) per second. Michell considered bigger and bigger stars. At some point, the mass of the star would be so big that the escape velocity was greater than the speed of light. What would happen if that were the case?

Michell speculated that gravity slows light down. If that happened he thought it would be possible to work out the mass of distant stars from their effect on light. But if there were bodies whose density is not less than the Sun, but with a diameter of more than 500 times that of the Sun, then light would not travel fast enough to escape and the star would be invisible.[5] There is an assumption here that light is influenced by gravity, something that there was no evidence of in Michell's time—not to mention the fundamentally flawed idea that light would travel slower away from a heavy star—and the idea faded into history.

The black hole (still unnamed as it wasn't until 1969 that American physicist John Wheeler came up with the term) reemerged wholesale and much better justified in 1916. The man behind this mathematical prediction was German physicist Karl Schwarzschild. In the First World War trenches, Schwarzschild somehow managed to find the time and concentration to work on Einstein's equations. At this point Einstein had only managed approximate solutions that predicted the effects on Mercury's orbit and the bending of light that came near the Sun.

Schwarzschild found the first exact solution of the equations for a very specific case. This was the gravitational pull of a perfectly spherical star. He would die within 4 months of his discovery, but his calculations lived on in their prediction of the existence of black holes. What Schwarzschild's solution of Einstein's equations showed is that there is a special radius called the Schwarzschild radius. To be precise this radius is $2GM/c^2$ where G is Newton's gravitational constant, M the mass of the star, and c the speed of light.

If a star somehow becomes smaller than this radius, specified by the star's mass, then at that distance from the star, a distance that would become known as the event horizon, bizarre things happen. Time (as seen from the outside) comes to a stop. Any photon of light that is emitted within that radius will never get out, because space-time is so strongly warped that it curves back in on itself, preventing the photon (or anything else) from escaping.

It might sound as if this means anyone passing through the event horizon of a black hole would experience strange phenomena—time coming to a stop, all of space-time being hugely distorted—but this is not the case. The stopping of time is a relativistic effect, as seen by an external observer. From the point of view of someone crossing the horizon, time would continue as usual.

As for the curvature of space-time, while this is sufficient to keep photons (and hence everything else) in, it does not have to be extreme at the event horizon. It is proportional to 1 over the mass of the star squared, so for very massive black holes, the curvature can be quite gentle at the event horizon. As the person entering the hole heads toward the heart of the black hole, known as the singularity, then the distortion *would* become extreme, tending to infinity, but it doesn't have to be noticeable to keep light in at the event horizon.

This singularity is the ultimate expression of general relativity, yet it is also the point at which, arguably, the whole theory falls apart. Because there is no reason why the star should stop collapsing, in theory at least, the black hole is a true point—dimensionless, with no size at all. This means you could approach it until the radius is 0. In practice you will have been shredded well before, but if you could make it, everything goes off the scale. The curvature of space-time and the gravitational pull is theoretically infinite. The mathematics collapses.

It ought to be pointed out that the singularity of a black hole is *not* a point in space at the center of the event horizon, although it is almost impossible to write about it without seeming to suggest this. It is a point in time. Once something passes the event horizon, it is headed for a point in time, the point of its arrival at the singularity. This is what makes the black hole inescapable. Any point in space could in principle be avoided, but a point in time can't. But even as a point in time, the singularity generates uncomfortable infinities.

Generally speaking, when physics predicts infinity in the real world, there is either something wrong or we are extending a theory beyond its capabilities. It is likely that under such conditions general relativity itself breaks down to be replaced by a different set of equations that don't go infinite. We can't tell in practice because even if we could directly observe a black hole, something that is yet to happen, there is no way for any information to come out past the event horizon, so we can never be sure how well the theory matches reality.

In a sense this is no surprise that general relativity breaks down because it is not a quantum theory. As such, it can't be expected to accurately predict the behavior of something very small—which is exactly what the hypothetical singularity at the heart of a black hole is. A singularity is a quantum phenomenon and could only be accurately described by a quantum theory of gravity (of which more in the next chapter).

Black holes are related to gravity in two ways. Their existence and behavior is predicted by general relativity, but also it is gravity that makes the black hole form if the star is big enough and has not got enough pressure to prevent the collapse. As we have seen, this will not happen for stars that are smaller than about twice the mass of the Sun. But that does not mean that smaller black holes can never be formed.

It is possible to make a black hole of any size if there is enough compression. A body smaller than twice the Sun in mass will not have enough gravitational compression for the process to begin, but much smaller masses could form a black hole if they were subject to immense levels of compression for other reasons. The amount of density required is phenomenal (though it may have been present soon after the Big Bang), but isn't prevented by the laws of physics.

So far we've seen black holes as very simple entities—the simplest things in the universe. No matter how complex the information that goes into a black hole—you could tip in all of Google's Web servers, for instance—the only thing that remains once it is in a basic black hole is the hole's mass (and by implication its size). Yet such a static entity seems highly unlikely. Pretty well everything in the universe spins around and it seems highly likely that a black hole would be spinning, too.

When you first think about it, it seems a little strange that everything spins, as if there is some kind of fundamental requirement. Planets spin, stars spin, galaxies spin . . . as far as we can tell the universe *doesn't* spin, but pretty well everything that has formed within it does. However we don't need a spin-obsessed genie, dashing around the universe giving everything a twist in order to expect this.

Think, for example, about the way a star forms. You've got a cloud of gas and dust that gradually coalesces under gravity. If that cloud had been perfectly uniform and spherical, and subject to no other influences, then it would have collapsed to form a star without a spin. But in practice it is lumpy and uneven; there is going to be more gravitational pull from one side than there is from another. The natural effect as the material comes together is that it is set moving in a rotating fashion and once the spin starts there is nothing to stop it.

Black holes usually form from stars. Those stars are spinning and there is no reason why they should stop doing so as they collapse into a black hole. We know that neutron stars spin because we can often detect the side effects of their rotation. They usually emit a stream of radiation from their magnetic poles, and as they spin around, this radiation sweeps around like the beams of a lighthouse. Just as we see a lighthouse flashing, we see such a neutron star pulsing—it's known as a pulsar.

The rate at which neutron stars rotate can be phenomenal. In the physical world various quantities are conserved. Mass/energy, for example. (We used to say energy, but because Einstein showed mass and energy are interchangeable, it is perfectly possible to create energy by losing a little mass—this is how nuclear power stations, nuclear bombs, and stars work.) One such quantity is angular momentum. Just as momentum (mass times velocity) is the amount of oomph with which an object moves in a straight line, angular momentum is the amount of oomph in a rotation.

This angular momentum depends on the ordinary momentum of each point in the object and the distance that point is from the center of rotation. If you take a rotating object and shrink it, then the distances from the center all reduce, so each bit of the shrinking object has to go faster. The result is that the rotation speeds up. (We're all familiar with this from spinning ice skaters, who suddenly spin faster when they bring their arms in beside their bodies.)

This gives the neutron star its sometimes remarkable speed. The first pulsar, discovered at the University of Cambridge's radio astronomy site by graduate student Jocelyn Bell, had a period lasting about a second. Others have been discovered since to have pulses that come less than 2 thousandths of a second apart, remarkably fast even for a small body. This object with a mass greater than the Sun can be spinning as much as 36,000 revolutions per minute.

There really is no reason to assume that black holes would not also spin, but Schwarzschild's solution of general relativity was only for static stars, which made for a simpler solution. It wasn't until 1963 that the New Zealand-born physicist Roy Kerr came up with a solution of Einstein's equations that allowed for black holes to rotate. Because of the rotation, the gravitomagnetic component of general relativity comes into play.

There will be a frame-dragging effect that means it's impossible not to be pulled around the hole with the black hole's spin. If an object did manage to stay still it would effectively be traveling faster than light, as light itself is dragged around the black hole by the frame-dragging effect. Such a black hole will also have a smaller event horizon than will a spin-free black hole. This is because the gravitomagnetic force will push objects (and photons) outward, meaning they can escape closer to the singularity than would otherwise be possible.

Black holes have two levels of existence. As first devised, they were simply mathematical structures—solutions to Einstein's equations of general relativity—that need not exist as a real object. Just because equations allow for a particular solution does not mean that it represents a physical object. The second level is the real entity—something to be observed out in space—and astronomers believe that they have detected a good number of black holes.

This may seem a contradictory idea. After all, unless we can detect them using light—or at least the whole spectrum of light that astronomers use from radio through to X-rays and gamma rays—we can never see a black hole. It is the ultimate black cat, sitting on a black wall, on a black night. But despite not emitting anything themselves, black holes are responsible for a considerable amount of radiation.

Firstly, there will often be an accretion disk. This is all the material around the black hole that is being pulled into it, typically from a neighboring star in a binary system. Although the black hole has no bigger gravitational pull at any particular distance than an ordinary star, material will travel much further before reaching the event horizon than it would before hitting the surface of a conventional star. That means it has more time to accelerate, plunging much faster toward the horizon, resulting in an unusual amount of radiation being given off.

The other component which is too small typically to be detected, but should still be there, is Hawking radiation. This is a quantum effect predicted by the English astrophysicist Stephen Hawking. Quantum theory suggests that "empty" space is a seething mass of virtual particles. Energy is constantly flipping into pairs of matter and antimatter particles, which exist for a tiny fraction of a second and then recombine as energy.

When this happens near a black hole, one of the particles will sometimes be sucked into the hole, leaving the other particle to escape. The black hole has turned the virtual particle pair into a single, real particle. This particle has mass; or to put it another way, it has energy. To keep the balance of conserved mass and energy, the particle that is sucked into the black hole has to effectively have negative mass and energy. The outcome is that Hawking radiation reduces the amount of mass in a black hole. Over time, if it isn't taking in too much ordinary matter from the outside, a black hole should evaporate.

Although technically nothing is coming out of the black hole, but rather something negative is going in, the result is that energy is leaking from the black hole through this quantum effect.

In black holes we see the clearest merger of gravity—the fundamental force behind their formation—and quantum theory, not only in Hawking radiation but in our understanding of what lies at the heart of the black hole. Yet whenever we speak of gravity and quantum theory in the same breath, we have a problem.

The other three forces of nature are understood as quantum effects. There is no reason why gravity, too, should not be a quantum force and if it could be approached this way, we could remove the infinities that plague black holes and the big bang. Yet general relativity remains a resolutely classical theory. It does not recognize quantum theory and the two cannot be merged. Something has to give.

CHAPTER NINE

ENTER THE QUANTUM

> *By far the most important consequence of the conceptual revolution brought about in physics by relativity and quantum theory lies not in such details as that meter sticks shorten when they move or that simultaneous position and momentum have no meaning, but in the insight that we have not been using our minds properly and that it is important to find out how to do so.*
>
> —"Quo Vadis" in *Science and the Modern Mind* (1971)
>
> Percy Williams Bridgman

Scientists are just as susceptible to hubris as anyone else. At the end of the nineteenth century, they thought they had physics pretty well sewn up. Within a few decades practically all of physics was undermined by the twin theories of quantum theory and relativity. It didn't mean that everything that had come before had to be thrown away. Newton's mechanics, for instance, was still practically useful, even though it was only an approximation to the more precise theory provided by relativity. But the foundations of physics were swept away to be replaced by these two new essentials.

And therein lies a problem. Quantum theory and relativity are not compatible. One (quantum theory) describes how reality works at the level of the very small—and it does so with astonishing accuracy. The other (general relativity) tells us how big things like galaxies and the universe as a whole behave. But despite the efforts of many, many physicists from Einstein onward, it has proved impossible to bring the two together. It's as if they operate in different dimensions that refuse to intersect.

Yet some combination of the two is necessary. Not only do we need to be able to pull the different fundamental forces into the same picture, there are aspects of the universe—like its origins in the big bang theory—that require the two to work together. What we are looking for is a quantum theory of gravity.

The field equations that underpin general relativity and that are tied up in that elegant little summary

$$G_{\mu\nu} + \Lambda g_{\mu\nu} = (8\pi G/c^4)\ T_{\mu\nu}$$

totally lack the essence of quantum theory. There is no quantization, no uncertainty, no probability. General relativity has a certainty about it that the real, quantum-based world can never have. This doesn't mean general relativity is "wrong" any more than Newtonian mechanics or basic electromagnetism is wrong. They work fine for the sort of objects we deal with in the everyday world. But when we get down to the level of quantum particles—or incredibly compact events like the big bang—these behaviors have to give way to the more detailed quantum version, where probability always plays a part.

Like Newtonian mechanics, general relativity deals not in probabilities, but absolute values. With relativity you can calculate the gravitational force generated by a mass to any level of accuracy—yet quantum theory tells us that we can't know the energy (for example) of a particle, and hence its mass, exactly. There is a mismatch. To come into line with the quantum theories that control everything else, there should be a quantum theory of gravity. A quantum particle with uncertain mass must have an uncertain gravitational pull.

There are two reasons that it's not enough to stick with "classical" general relativity. First, it simply can't be applied to quantum particles. And second, as we have seen, there are phenomena in the universe (other than particles) that have inherently quantum natures, requiring quantum gravity if we are to understand them properly.

Take the big bang. Go far enough back in time, close enough to the big bang itself, and you are dealing with a purely quantum phenomenon. Yet it's one where gravity is an essential component. Without being able to combine the two it is impossible to say anything meaningful about the big bang itself.

Just think about that statement. Next time you see a TV special on which a celebrity scientist is pontificating about the big bang as if it were solid fact, you will know better. The big bang is an idea that emerges from a theory that doesn't work at that scale.

Without quantum gravity, we can't say what happened at the big bang or even if it existed at all. The only certainty is that our current scientific models fail. Entirely. The same goes for the singularities at the heart of black holes, not to mention some of the more exotic concepts of cosmology like negative energy and wormholes.

At the moment we don't have a clear quantum theory of gravity. Faced with a reality that exposes limits in our current theories we have to go further, but as yet the only approach has been to explore more and more complex mathematics. Often the math itself will throw up many possibilities. Until we have predictions that can be tested by observation and experiment—something that doesn't yet apply to any quantum theory of gravity—we are limited to making mathematical rather than physical decisions on what makes a good theory. Which doesn't make for the best science.

One criterion that mathematicians can use is consistency—ensuring that a mathematical structure does not produce different results that contradict each other. Another, more controversial, is beauty. This is a subjective criterion that mathematicians have difficulty providing good descriptions to support. They consider some equations, some bodies of mathematics to be more "beautiful" than others, where part of that beauty is a sort of simplicity (though the math may be furiously complex to actually carry out).

The danger here is twofold. One is that not all mathematicians will agree on the beauty or otherwise of a collection of equations.

The other is that mathematical beauty is not necessarily the result that nature provides. Although natural laws are often parsimonious—for example requiring the least time or effort to undertake a particular action—they aren't necessarily based on the most elegant option. (You only have to take a look at biology to see this is the case.) It is entirely possible for mathematicians, with no grounding in real observation or experiment, to come up with a beautiful castle in the air that has no value to science.

Don't get me wrong, there is no problem with getting the math right first and then seeing if the science fits your theory. This happened in the late 1920s when the unfairly obscure English physicist Paul Dirac found a way to combine quantum theory and special relativity. Up to that point, quantum theory had been based on Newton's ideas of time and space. Dirac managed to modify Schrödinger's equation describing the behavior of quantum particles so that it was relativistic.

This did more than just tweak the results. It brought a new component into the equation. In a circumstance where Schrödinger's equation had one solution, Dirac's equation has two. The consequences of these extra values was to predict the existence of a different form of matter—antimatter. At this point in time, antimatter had never been observed, nor was anyone looking for it. But observation soon proved Dirac right, vindicating his elegant mathematics.

Here was an enhancement to science that emerged directly from using better math in the scientific model of reality. This was mathematical beauty indeed. But Dirac had not achieved what we are looking for in merging general relativity and quantum theory. He was only working with special relativity, the correction of space, time, and energy to deal with movement. We need something that goes many steps further. General relativity's mathematics are vastly more complex than special relativity, and unlike Dirac's equations, all the attempts at producing an effective model of quantum gravity make no testable predictions.

At the moment there are two main theories competing to supply the answers for quantum gravity. By far the most effort has gone into string theory (and its younger brother M-theory), though as time passes, loop quantum gravity is gaining increasing interest. These aren't the only possible approaches. At the moment the field is wide open, and because it is being driven more by mathematics than physics, so arbitrary that there could in principle be many, many different options.

There is, for example, twistor theory. This exotic-sounding model, developed by English physicist Roger Penrose, starts from the difficulties of applying the smooth differential equations (which assume that reality is a continuum), at the heart of general relativity, to the sort of discontinuous space-time that may be implied by quantum gravity.

After all, it is entirely likely that the attempt to pin down a quantum approach to gravity will produce a space-time that is itself quantized: A space-time that is digital rather than analog, broken into "atoms." This feels like an attractive idea, because it does away with all the problems that emerge from particles (or singularities) being dimensionless points.

Take, for example, the singularity, thought to be at the heart of a black hole. This is supposed to have totally collapsed, to have no spatial size whatsoever. But the gravitational force grows with the inverse square of the radius. As you get near to the singularity, the gravitational force should shoot up toward infinity. At the singularity itself it should *be* infinite—and physics as we know it breaks down.

As loop quantum gravity proponent Martin Bojowald points out: "... infinity as the result of a physical theory simply means that the theory has been stretched beyond its limits; its equations lose all meaning at such a place. In the case of the big bang model, one should not misunderstand the breakdown of equations as the prediction of a beginning of the world, even though it is often presented in this way. A point in time at which a mathematical equation results in an infinite value is not the beginning (nor the end) of time; it is rather a place where the theory shows its limitations."[1]

If, however, space-time is quantized, broken down into minimum-sized lumps below, which it is not possible to go, then nothing, not even a singularity, can be truly said to be without dimension, as there is no meaning for anything smaller than one of those lumps. The best guess for the dimensions of such quanta of space is the Planck distance. This is a tiny distance constructed from fundamental constants.

If the Planck length is the quantum size of the space part of space-time, those grains of reality are around 1.6×10^{-35} meters (6.3×10^{-34} inches) on each side. That's a ridiculously small distance: 10^{-35} means 1 divided by 1 followed by 35 zeros. There are 0.16 billion, billion, billion, billion such units in a meter. Similarly, in quantum space-time, time would be limited to the time light takes to cross such a distance, around 5.4×10^{-44} seconds.

Of itself, making space-time quantum seems a simplification, because there are less points in space-time to deal with. But in mathematical terms quantization of space-time has the potential to release a monster. Twistor theory is one approach to dealing with this. Instead of working in four normal dimensions of space and time it works in four complex dimensions.

This is "complex" in the mathematical sense, rather than simply complicated. A complex number combines a real component and an imaginary component, where the imaginary part is the square root of a negative number. Imaginary numbers date back in a crude form to the sixteenth century and reflect a simple problem that arises from the fact that multiplying two positive numbers together provides a positive number (for instance $3 \times 2 = 6$), and multiplying two negative numbers together also gives a positive result (so $-3 \times -2 = 6$).

That isn't a problem until you start thinking about square roots. It's easy enough to see that $\sqrt{4}$, the square root of 4, is 2 or −2, because $2 \times 2 = 4$ and $-2 \times -2 = 4$. But what is $\sqrt{-4}$? What number multiplied by itself produces a negative number? It can't be a positive number—the result would be positive. And it can't be negative—the number would be positive again. If there is a square root of a negative number, it must be something else, something special. Quite arbitrarily it was decided that there could be such a thing. The square root of −1 was given the symbol i, making $\sqrt{-4}$ become 2i, twice the square root of −1.

The inspiration that brought *i* into the toolkit of physicists and engineers was the realization that you could use imaginary numbers to extend the number line into a new dimension. Think of a traditional number line, stretching off horizontally to the unimaginably large negative at the left, passing through 0 in the middle and heading off toward infinity at the right. Now put a second line at right angles to this, passing through 0, like the second axis of a chart. Call this the imaginary number line.

As you move upward from 0 you are going through bigger and bigger imaginary numbers: i, 2i, etc. Below the conventional number line you have negative imaginary numbers. This was the contribution to the development of imaginary numbers made by the nineteenth-century German mathematician Johann Carl Friedrich Gauss, and it was the final step needed to transform imaginary numbers into the powerful tools they are today. Any point on that number plane is a complex number—a combination of a real number on the horizontal number line and an imaginary number on the vertical number line.

In effect, twistors double the information that is contained in a location in space-time, while still maintaining the four key dimensions. The traditional approach to working with imaginary numbers is to make sure they disappear before you get to an answer. They can play a useful part in the calculation, but you mustn't have an imaginary outcome (for example an object that is 3i meters high). In twistor space, the complexity is real, though in translating the outcomes to normal space, the imaginary components disappear.

Another approach that is being examined to see how it might help with an understanding of quantum gravity is noncommutative geometry. Some mathematicians have suggested that movement in such quantum space-time is no longer the simple matter of jumping from point to point in a multidimensional matrix of space and time, but becomes a shift in quantum symmetry.

Symmetry lies at the heart of much modern physics. Its importance began with the work of the first great female mathematician, Emmy Noether. Born in Bavaria, Germany, in 1882, Noether was the daughter of a mathematician and was a rare student of mathematics at a time when few women even went to university.

At the start of her career she could not get a paid position, but lectured unpaid, and even when she won a post in the mathematics department at the University of Göttingen, under the aegis of the great mathematician David Hilbert, she was forced to lecture under Hilbert's name for several years while the powers-that-be argued over the rights and wrongs of having a woman teach the students.

Noether's masterwork (at least as far as physics is concerned) was on symmetry. Symmetry is a property of practically anything built on repeating structures. The most familiar form is probably mirror symmetry, where one half of the object or space is the equivalent of the other half reflected in a mirror. So, for instance, a square has mirror symmetry along both the lines that divide it in half down the middle and along its diagonals. A circle has mirror symmetry on any diameter.

Then there is rotational symmetry—where you rotate the object or space through a number of degrees and get something indistinguishable from the original. Our square has rotational symmetry when turned through any combination of 90 degrees about the middle, while the circle can turn through any angle and remain symmetrical.

Perhaps less obvious, but more general, is translational symmetry. This is when an object or space is moved either in space or time. A simple shape has translational symmetry in time, because it doesn't change with time. But such shapes don't have translational symmetry in space. Admittedly the shape itself doesn't change if you move it sideways, but the whole space including the shape does. Imagine you've a square at the top of a sheet of paper and a second sheet with the square shifted halfway down the page. You can easily distinguish between the two sheets of paper, but all that has happened is a translational move of the shape.

Translational symmetry often occurs in regular patterns. For example an infinite surface covered in squares has translational symmetry when you move it either horizontally or vertically by the size of one square—there is no way of distinguishing between the two copies of the space.

There are other, more complex symmetries, plus any of these symmetries can be combined, and they provide a rich playground for abstract mathematics. But Emmy Noether proved that they were more than mathematicians' playthings. Noether's theorem (as it is called) proved valuable for physicists because it links symmetries and conservation laws.

Conservation is one of the key fundamentals of physics, meaning that there are certain things that can't just appear or disappear. So, for instance, we have already seen the conservation of energy (or more properly mass/energy). Noether showed that symmetry in a physical system resulted in the conservation laws. So, for example, if something is symmetrical through time, then Noether proved that energy must be conserved. If it is symmetrical through translations in space, momentum (mass times velocity) is conserved. And so on.

Symmetry lies at the heart of the current ideas on the fundamental particles in physics. However if space-time is quantized, then we end up dealing with quantum symmetry, a less intuitive concept, that to all intents and purposes makes movement in space-time something that isn't amenable to conventional mathematics, but requires a more complex approach where the geometry is noncommutative.

We are used to both commutative and noncommutative operations in ordinary arithmetic. If something is commutative, the order in which we put the thing operated on doesn't change the outcome. So 2+3 and 3+2 are both the same. Similarly with 2×3 or 3×2. But when arithmetic gets more complicated, things aren't so obvious. Is $2+3\times4$ the same as $3+2\times4$? It depends: $(2+3)\times4$ is the same as $(3+2)\times4$, but $2+(3\times4)$ is not the same as $3+(2\times4)$.

If geometry becomes noncommutative, the order in which we make movements in space and time have different outcomes. It is no longer possible to have a clear link between the math describing what is happening and the phenomena observed. The math almost operates in isolation and would produce the right outcome, but there would be no intuitive link between the two.

A similar thing happened in the early days of quantum theory. As we have seen, German physicist Werner Heisenberg and others found a way to use manipulations of matrices—abstract collections of numbers—that would predict what would happen to quantum particles. But there was no intuitive connection between this "matrix mechanics" and any picture of reality.

Matrix mechanics was later proved to be entirely compatible with the wave equation approach to quantum theory developed by another German physicist, Erwin Schrödinger, which had an imaginable "real world" meaning and the two approaches were merged. Neither was technically better than the other, but the matrix approach was isolated from our understanding of the behavior of particles. It was simply a series of numbers, which when manipulated produced the right predictions. The approach currently needed to work with quantum space-time using noncommutative geometry is similarly nonintuitive.

The implications of quantum space-time are still not totally agreed, but according to some scientists there is a way that we could distinguish between space-time that is quantized and space-time that is continuous. Some models of quantum space-time predict that the speed of light should depend on its frequency—blue light, for example, should be a very small amount slower than red light. This makes it, at least in principle, a proposition that could be tested. There have been some results that hint at such a variation, but others suggest it does not exist.[2]

For the moment, noncommutative geometry is something that keeps a number of mathematicians and theoretical physicists busy, but remains an uncertain adjunct to quantum gravity. It might prove the best line of attack, as may twistor theory—but for the moment string theory and loop quantum gravity continue to get the most attention.

String theory was born from an attempt to get a better handle on the burgeoning collection of particles that makes up the "standard model" of particle physics. The standard model reflects the outcome of a period in the 1960s and 1970s when it seemed that the particle zoo was getting out of hand as more and more particles were discovered in collider experiments and cosmic rays.

In the standard model there are twelve particles related to matter (the six quarks, leptons, which include the electron, the muon, and the tau, and three neutrinos), plus five bosons, the carrier particles of forces—the photon for electromagnetism, W+, W–, and Zo for the weak nuclear force, and the gluon for the strong nuclear force. There are then a few oddities—the graviton as another boson, assuming gravity *is* quantized, and the hypothetical Higgs particle that is responsible for some mass.[3]

Then, to make things even more messy, all the matter-related particles have antimatter equivalents, so despite the apparent simplification of the standard model (because a lot of the familiar particles are built from quarks), it is still a complex collection with around thirty particles on its books. String theory slashes through this ungainly assembly of ugly relatives at a stroke, leaving a single fundamental building block. That sounds very attractive. How could things be simpler than that?

The idea is that each one of those particles is not what it seems, but is simply a different way of looking at a more fundamental entity. Each particle hides a chameleon object that manages to give the impression of being all the others through the way it moves. This object is called a string. Not a literal, physical piece of string, but a one-dimensional object that through its different vibrations, both as closed loops and open strings, produces the different particles.

As we have seen, mathematicians and mathematical physicists like "elegance" and "beauty" in their math. As a concept, strings have an elegance that is easy to appreciate without any mathematics to back it up. There is a neat simplicity that makes the idea attractive and easy to get hold of. Just as the string on a violin or guitar can produce different notes if it vibrates in different ways, so the particle string can produce the different particles from its various modes of vibration.

The price of this elegant simplicity is twofold. The mathematics to support it is fiendishly complex—and it only works if there are nine spatial dimensions plus time. Those nine dimensions present more than just a conceptual problem of how to imagine so many dimensions. They lead to the biggest issue facing string theory enthusiasts. They result in what has been called a rich set of solutions.

In truth, "rich" isn't the half of it. There are more possible solutions to the equations of string theory than there are protons in the universe. Ridiculously many. And the mathematics that brings string theory into being provides no mechanism for choosing between them. As they stand, the equations predict anything and nothing. They may delight mathematicians, but they are useless to a practical physicist.

The search for a unifying theory for gravity and the other forces of nature has been described as the hunt for a "theory of everything." And as physicist Martin Bojowald has pointed out, string theory as it stands is a literal theory of everything because everything—and anything—can happen in it.[4]

It might seem obvious that string theory is wrong because there aren't nine spatial dimensions. According to the theory, as well as the three we are familiar with we would expect a fourth dimension at right angles to all of those, a fifth at right angles to the first four, and so on. String theorists accept that we don't see any more than the three dimensions and explain this by suggesting that the other spatial dimensions are curled up so small that we can't detect them.

Quite separate from the many billions of solutions, there are five major variants of string theory. These, at least, have been combined into a single approach called M-theory (according to those involved, no one is quite sure what the M stands for, but this suggests to me that M means marketing). M-theory produces an even richer picture than the various string theories it combines, requiring ten spatial dimensions to fully describe reality. Within its eleven-dimensional vision of space-time, the basic unit is a brane, which can have a one-dimensional form as a string, but also can have many more dimensions.

In M-theory, our universe is a three-dimensional brane floating in that ten-dimensional space, which allows for some interesting possible theories for the nature of the universe as a whole. Just like string theory, M-theory provides vast numbers of solutions and no testable predictions. As British physicist Paul Davies has commented: "[the complexity and lack of predictions] leaves string/M theorists without much of a reality check. Where this enterprise will end is anybody's guess. Maybe string/M theorists really have stumbled upon the Holy Grail of science, in which case one day they might be able to tell the rest of us how it works. Or maybe they are all away in Never-Never Land."[5]

In some respects the main challenger to string theory and M-theory is much simpler, because it does not involve the complexity of ten- or eleven-dimensional space-time. But there is a price to pay. Loop quantum gravity lacks the simple image of the string as a fundamental unit that lies behind all fundamental particles. Instead we have to change our understanding of reality to a more abstract view.

Traditional quantum theory deals with quantum particles like an atom of matter or a photon of light. These are independent objects within space-time, though they can be considered collectively by using quantum field theory, the mechanism quantum mechanics uses to deal with a complex system with many particles involved. In loop quantum gravity, space and time are quantized, broken into "digital" units, as we have already seen in some of the less well-supported theories. Space becomes a construct of spatial atoms, not physical objects in their own right but the logical components of space.

These atoms of space have properties that act as if they were a loop of one-dimensional material—this is why the theory is called loop quantum gravity.[6] Just like matter atoms, atoms of space can be excited by having extra energy added to them. Matter atoms are typically excited either by heating or when they absorb the energy of a photon. Unlike matter, spatial atoms grow as a result of their excitation, and can split, rather like a biological cell, resulting in space that expands like a rapidly dividing embryo.

There is, then, a fundamental difference in approach between loop quantum gravity and its main competitor. String theory, which originated from particle physics, takes quantum theory and lays it on top of general relativity's space-time. In the process it moves away from four-dimensional space-time to something more complex, imposed by the mathematics. Loop quantum gravity applies quantum theory to space-time itself, but does not require this extension to general relativity. In that sense it is a more fundamental approach, though this does not necessarily mean that is better or more capable of matching reality.

Breaking space-time down into atoms seems simple enough, but quantum theory is never straightforward. (Loop quantum gravity works initially on space alone then adds time in, but space-time always has to be the ultimate goal.) One of the more dramatic implications of conventional quantum theory is the uncertainty principle. This says that it is impossible to know both the position and momentum of a quantum particle with perfect accuracy. The more you know about a particle's position, the less you can know about its momentum and vice versa.

Once space-time itself is quantized, that, too, has to be subject to an uncertainty principle. So, for example, an area of space and its curvature would have the same kind of uncertainty relationship. The more accurately you know the size of an area, the less certain you can be about its curvature. The more detailed your information on the curvature of a part of space, the less knowledge you can have about its area.

From the loop quantum gravity viewpoint, the loops—the atoms of space-time—create the geometry with which space is built. Where images of general relativity's continuous space-time were typically based on sheets of rubber, loop quantum gravity's space-time is represented as a weave of loops, where surfaces and volumes are created by the intersection of these curvy space-time atoms.

There is a conceptual hurdle to be got over here. If we imagine a set of loops creating space-time, we inevitably think of these loops existing within space and time, spreading through the space-time continuum to give it substance. But this picture isn't low level enough. The loops aren't *in* space-time, they *form* space-time (or more accurately space, as time remains something of a bolt-on in loop quantum gravity). If you had a region with no loops there would be no space and time in it. It would be truly empty in a way that the vacuum of space isn't. Without loops nothing can exist, no light can travel—a loop-free region is true nothingness.

Even more so than the strings of string theory, the loops of loop quantum gravity are not entities in the normal sense. A more accurate version of a string would be whatever a particle really is, but having some of the characteristics of a one-dimensional vibrating string. Loop quantum gravity's loops are mathematical constructs that enable the construction of space. We don't know that anything exists outside the math. If they are real, they are ridiculously small, with around a googol (10^{100}) in a cubic meter.

To make matters worse for those trying to visualize the loops, we need to remember that they are quantum entities. When atomic structure was first discovered, an electron in an atom was portrayed like a planet whizzing around the Sun (it often still is in graphics). But being a quantum particle, the position of an electron is uncertain, and it is much better to imagine it as a fuzzy cloud of probability. Similarly the loops in loop quantum gravity and the fabric of space they form are not clear, well-positioned links like a chain mail, but fuzzy probabilistic entities, constantly on the move.

Although the first attempts at loop quantum gravity concentrated on space, it clearly needed to handle space-time. As different loops are added, removed, and interact in a quantum fashion, so, too, does time when seen through the loop quantum gravity perspective. Just as there is no space in between the loops, there is no time between the steps of changes to loops. Time does not flow steadily but progresses in fixed increments.

There is a huge implication of this quantization of time when the clock of the universe is run backward and we approach the big bang. Because of the lack of "in-between times" it is perfectly possible to jump back from the point immediately after the big bang to the point immediately before. If the big bang is assigned to a nonexistent interstitial point in the fabric of space-time, loop quantum gravity will slip back in time through the big bang into the pre-big bang universe.

In this theory, the big bang is no longer a physical impossibility where infinity comes on the scene, it is a nonexistent transition between two atoms of space-time. There is no big bang. It is a non-space at a non-time. It is likely in such a picture there was an earlier contracting universe that went through a symmetry reflection at what should no longer be called the big bang, but rather the big bounce.

The ability to prevent the creation of a singularity at the big bang, or in a black hole, is just a side effect of the nature of quantum time. In traditional general relativity (it comes to something when the infamously complex general relativity is "traditional"), there is no stopping the force of gravity. As the object gets smaller and smaller the gravitational force goes through the roof, producing an unstoppably vast attraction. There simply is nothing that can resist it. It has to crush the object all the way to zero size, with its energy density soaring off to reach infinity.

Once time is split into quantum "atoms," there is a limit to how much energy can be crammed into a unit of space-time. That breakup of reality into discrete segments acts as a kind of barrier; when its energy capacity is full an object simply cannot contract anymore because there is nowhere for the energy to go.

A useful peculiarity of the way loop quantum gravity works takes out a potential difficulty with omitting the big bang. It is all very well to say that the big bang sits in the nonexistent gap between atoms of time, but what if things were different? Why should the big bang be positioned so conveniently on the timeline as to fall in a gap? Remarkably, this behavior falls out of the math. As loop quantum gravity expert Martin Bojowald puts it, likening the atoms of quantum time to frames in a movie:[7]

> It turns out that the elementary dynamics of loop quantum cosmology, which already gave rise to the discreteness in the form of having only finitely many frames, automatically takes out the one that corresponds to [the big bang]. Even if one tries to arrange the time steps in such a way that time zero, when the full collapse is reached, would be included as a frame, that frame is automatically detached from the rest and does not cause a singularity.

In analyzing the nature of quantum space, supporters of loop quantum gravity have identified two possible states for any particular volume. In the material world we are used to the idea that every matter particle has its antimatter equivalent with a reversed charge. Scientists at CERN (not using the Large Hadron Collider as is often suggested) have produced anti-atoms of hydrogen, and in principle you could have an antimatter version of anything from you to the Earth. Loop quantum gravity suggests that space, too, has a kind of antispace.

As space is neutral as far as charge goes, we are not looking at a charge to distinguish between, say, positive space and negative space. Instead the two types of space form a multidimensional mirror image. The predicted antispace (my term) is simply existing space turned inside out. This is reminiscent of a wonderful scene in the movie *Galaxy Quest.* The premise of the movie is that technologically advanced aliens pick up the broadcast of a TV show that is suspiciously similar to *Star Trek* and, believing it to be true, produce an identical copy of the fictional space craft.

In the scene I have in mind, the actor who plays the ship's engineer in the TV show is trying to operate a real matter transmitting transporter. He gets the controls wrong and the creature he transports (luckily not one of the crew) arrives inside out. It's not a pretty sight. However, when we talk about space being turned inside out, this isn't the image we should be looking for. Inside out "antispace" would not feature horribly distorted creatures.

For its occupants, an inverted space would seem no different to the way our space does to us.

Unlike matter and antimatter which coexist (if rather uncomfortably) in the same universe, the suggestion is that space and antispace exist in different universes. It could be that the big bang was simply an inversion point where the universe turned inside out from its antispace past to its space future.

One interesting prediction loop quantum gravity *does* make is that gravity should be left-handed. Specifically, it tells us that the graviton particles that carry the gravitational force should have their spin aligned with the direction of motion (somewhat arbitrarily defined as making them left-handed). In principle gravitons could come with either spin direction, but loop quantum gravity requires the left-handed version.[8]

This isn't very useful unless you can detect the spin of gravitons (which themselves are yet to be discovered). João Magueijo and Dionigi Benincasa of Imperial College, London, have suggested that there could be an indication in the cosmic microwave background, the so-called afterglow of the big bang. The Imperial team believes that if gravitons are all left-handed this should leave a particular pattern in the way the photons that make up the cosmic microwave background are polarized. An analysis of data from the Planck space telescope should clarify whether this is possible by 2013.

Of itself such a discovery doesn't rule out many of the other theories of quantum gravity, as most would allow for left-handed gravitons, though should they prove all to be right-handed it would be worrying for loop quantum gravity. Having said that, Magueijo and Benincasa's theory piles assumption on assumption, requiring gravity waves of a particular kind to be emitted during the inflation of the universe, so it would hardly make a definitive proof either way.[9]

Loop quantum gravity probably has more support at the time of writing than any other theory of quantum gravity apart from string theory, however there is a pretender in the wings, developed by Peter Horava, a Czech physicist working at the Berkeley Center for Theoretical Physics. Horava has long been a supporter of string theory, but has recently challenged our fundamental understanding of gravity, and specifically the existence of space-time, the four-dimensional environment that underpins both special and general relativity.

Horava believes that space and time should be treated separately and, since publication of his basic ideas in 2009, has continued to develop an approach that challenges the standard way of looking at gravitation. As we have seen, loop quantum gravity considers space first and adds in time, but Horava is proposing a more fundamental split. Rather than try to combine general relativity and quantum theory, Horava wants to dismantle general relativity and start again in a way that removes many of the problems facing anyone trying to produce a unified theory.[10]

Horava's inspiration for this new way of thinking about relativity was graphene. Graphene is an incredibly thin layer of carbon in graphite form—pencil lead—that is just one atom thick and that displays some fascinating properties. So impressive is graphene that the investigators working on it at the University of Manchester won the 2010 Nobel Prize in Physics.

Despite some very low technology in its investigation—the graphene sheets used in the Manchester experiments were produced using regular adhesive tape to separate off an atomic layer—the properties of graphene are truly exciting. Although transparent, it is a great conductor. The electrons in the material zip around as freely as in a metal. But Horava's inspiration was from graphene that is supercooled to just a few degrees above absolute zero.

Generally speaking the electrons conducting electricity move at a leisurely pace—it is an electromagnetic wave that carries signals at high speed down a wire. But in the supercooled graphene, the electrons were able to get up to sizeable percentages of the speed of light, fast enough for special relativity to come into play. Yet uniquely, in the two-dimensional world of graphene, it seems that the rules of special relativity don't always behave as expected.

We normally say that pinning down the speed of light in relativity meant that other factors previously constant are loosened up—specifically time slows down and length decreases in the direction of travel. This tit-for-tat change, a form of symmetry, seems not to apply in the same way in graphene in all circumstances. When the electrons were first moving at relativistic speeds the symmetry seemed to be broken, with the normal special relativity relationship returning as they were further cooled.

Horava wondered if there might be a similar breakdown of

special relativity in the early stages of the universe, with its symmetry restored as the universe cooled. He removed this symmetry from Einstein's general relativity equations and the result was a form of general relativity that *is* compatible with quantum theory. By also adding in a specific preferred direction to time (general relativity doesn't care which way time flows), he developed a new working theory of quantum gravitation.

The price was the breaking up of space-time, which is fundamental to both special and general relativity, but which can't hold together if the symmetry doesn't apply. This means that rather than distorting together, space and time can be warped separately in different ways. But this apparent complexity in understanding of the universe results in much simpler mathematics than that required by string theory or loop quantum gravity.

What's more, so-called Horava Gravity has other benefits. Cosmologists have had to devise dark matter, undetectable heavy material out in space, to explain why galaxies don't behave the way general relativity predicts. They have been observed to rotate faster than was expected, producing the inference that there was extra mass holding them together. But if general relativity is not quite right, the distinction between its results and the predictions of Horava's theory could be enough to explain away the movement of the galaxies without any mysterious undiscoverable mass.

Horava Gravity is not as polished a theory as string theory or loop quantum gravity. There is quite a leap to be made from the behavior of graphene to that of the universe. And special relativity seems to work just fine at room temperature, not just at temperatures near absolute zero. Nonetheless, Horava's theory is an interesting one that has received a lot of attention at the moment, demonstrating that the solution to quantum gravity may require changes to something as fundamental as general relativity, rather than simply providing the glue between relativity and quantum theory.

At risk of overloading you with theories, I would like to explore one more because, like Horava's it demonstrates that the problem with string theory and loop quantum gravity is that those working on them might be trying too hard to preserve old ways of thinking. Back in the 1990s a man with a humble physics degree working in industry got the quantum gravity bug. So fascinated was he by the possibility that he went back to university and ended up writing his Ph.D. thesis on a revolutionary way to combine quantum theory and general relativity.

This was in 1996. At the time of writing, that man, Mark Hadley, is still developing his theory further in the post he gained at the University of Warwick in England.[11] There haven't been many others who have taken up the theory. Hadley lacks Horava's contacts. But there is no doubt that Hadley's theory would benefit from further examination. Hadley's approach turns the usual attempts to unify gravity and quantum theory (and Horava's in particular) on their heads. Rather than modify general relativity to fit quantum theory, it derives a version of quantum theory from the nature of gravitation.

In Hadley's theory, quantum particles are produced by gravity, artifacts of the gravitational field.[12] From this viewpoint, a particle is an intensely tight warp in space and time. This is something most general relativity theorists from Einstein onward have been wary of considering. The reason is simple: if you warp time enough, you end up with the capability to perform time travel. Tightly warped time puts the usual causal connections that if A causes B, A has to come before B at risk. But this, Hadley suggests, is an essential if you are to generate the probability-based nature of quantum theory from the clockwork certainty of general relativity.

When we look at quantum particles it seems puzzling that we can't usually specify exactly where they are, that they can be in more than one place at a time as a probabilistic wave, rather than occupying a single location. But Hadley's theory suggests this is inevitable because of the way they are formed. The tiny knot of time that he believes creates a particle means that it can be influenced by both the past and the future, and with this one change to reality it is hardly surprising if the quantum particle has these strange behaviors.

Of course Hadley's theory does not depend on hand-waving similarity, but on math that demonstrates how the wave equations that specify the properties of quantum particles can be derived from a knot in space-time's connections in the past and future. But there is a very attractive feeling that this explanation makes sense, particularly when you look at some of the specific peculiarities of quantum theory.

Take beam splitters. The simplest form of beam splitter is a house window at night. Look at the window from the inside of a lighted room and you see the room's interior reflected. But go outside and you can clearly see in. Some light is reflected by the glass, some passes through. At the level of photons, the quantum particles of light, we can't say if any particular particle will pass through or be reflected, but we can specify the probability of this happening.

Now the bizarre thing about beam splitters is that the number of photons that pass into the glass from the room is influenced by the thickness of the glass. Change the thickness and the percentage that are reflected at the surface changes. Somehow those photons know how thick the glass is. Quantum theory explains this by telling us that the photons are not located at a point but are spread out as a probability wave that enables them to "be aware" of the thickness of the glass—but this is an exact parallel of Hadley's "view into the future" of quantum particles. For a particle that is influenced by the future, it is no problem to know whether to reflect or pass through depending on the thickness of the glass.

Hadley's theory has some serious problems as it currently stands. If particles were truly tiny time machines we might expect the apparently ordered universe to fall apart, something we clearly don't observe. Hadley has suggested this might be because the space-time knot forms a kind of event horizon, like a black hole does, that makes it impossible to see into the future from the outside.

There are also serious difficulties finding suitable solutions to

Einstein's equations of general relativity to allow for these space-time knots to form as the quantum particles we observe. But even if it is fatally flawed, Hadley's idea is a useful prompt that it may be necessary to think very differently about quantum theory and general relativity if the two are to be united.

As we have seen, several quantum approaches to gravity have been proposed, but as yet none is wholly satisfactory. More to the point, as yet there are no observations or experiments that offer a mechanism for checking the theories. We have a level of understanding that could be considered on a par with announcing that "the universe is run by an invisible, undetectable dragon." I can say that all the behavior of the universe, including gravity is explained by a super-powerful, omnipresent dragon that makes everything happen the way it does. Because there is no way to prove or disprove my assertion (remember, this dragon is undetectable) it is useless as science, *even if it is true.*

Those last five words of the statement provide an assertion that many scientists don't realize is part of the baggage of untestable theories. It is not uncommon for science to dismiss the existence of God, where what is really meant is that there is no way to prove or disprove the existence of God, so from a scientific viewpoint it is best to leave a deity out of the picture. Science doesn't provide the mechanism for taking an anti-religious stance, nor, for that matter, a pro-religious stance.

At the moment, the best theories of quantum gravity like string theory and loop quantum gravity are in a similar position to a religious belief. We have a collection of concepts, often in this case arising out of complex mathematics, but no observations to support or detract from a theory. This doesn't mean the observations and experiments won't come, but until they do, we shouldn't put too much weight behind any particular theory. Which is distressing for those who have dedicated their entire career to an approach like string theory.

This doesn't mean that we can't suggest possibilities for seeing quantum gravity at work. It is possible, for example, that evidence of quantum gravitation will come from the period in the infanthood of the universe when the lightest of the elements were first formed.[13] Most of the familiar elements that make up the Earth and our bodies were produced in stars, but hydrogen, helium, and some lithium have been around since the early universe cooled enough for matter to form.

We would expect at this stage that there would be subtle differences between the action of classical general relativity and the modifications to the theory that must be required to allow for quantum gravity. This would alter the exact proportions of those early-formed atoms, depending on the nature of quantum gravity. Here is something that can already be measured and could be taken to greater accuracy that could show us the influences of quantum gravity, provided the theories can ever be persuaded to produce predictions to test against observation.

Another possibility, to see whether or not space-time is truly quantized, as we have already seen, is to try to detect the impact of the granularity of space-time on the light that travels through it. White light passes through empty space unchanged, but when it hits a medium like glass, the atoms in the medium interfere with its steady progress. Photons of different energy will have different levels of interaction, which means that the medium tends to split white light into its constituent colors. This is how we get rainbows.

If there are atoms of space as loop quantum gravity and some other theories suggest, then we would expect there to be slight differences in the way photons of different energy propagate. These differences would be very small, so small that it would be necessary to have an extremely intense burst of light coming from a vast distance before anything could be noticed. Luckily such cosmic fireworks do exist in the form of gamma-ray bursts.

These may be caused by collapsing stars, colliding neutron stars, or evaporating black holes—no one is quite sure how gamma-ray bursts originate, but they exist and they are extremely intense. Imagine an outpouring of the entire lifetime energy capacity of the Sun in anything between an hour and a fraction of a second—that's a gamma-ray burst. They are so bright that they can be seen at vast distances, in which case the light has been traveling for a long time. (It's just as well that they are mostly at vast distances. A gamma-ray burst from a source within a few thousand light-years would devastate life on Earth.)

This makes gamma-ray bursts an ideal mechanism for trying to detect the atomic structure of space. What's more, the high energy of the photons gives their wavelike properties very short wavelengths, making them more susceptible to influence by the extremely small granularity that quantum space would have.

The highest energy gamma waves should travel slower if space is quantized.

The first hint that there might be data to support this theory came on June 30, 2005, when the Major Atmospheric Gamma-ray Imaging Cherenkov Telescope (a name that really could only be so clumsy because it was retrofitted to spell out "MAGIC") on La Palma in the Canary Islands off North Africa, detected a burst of energy from a galaxy called Markarian 501, around 500 million light-years from Earth.[14]

In the stream of photons arriving after their 500-million-year journey, the highest energy group were around 4 minutes behind the least powerful ones. Inevitably such a phenomenon could have a range of causes. It could simply be that the way the burst originated, with lower energy light the first to emerge, followed by the more powerful photons. But equally this detection event—and a string of similar events that followed—could be the first signs that space really is granular.

There is a limit to what is possible from an Earthbound observatory, but there is already a satellite specializing in searching out gamma-ray bursts, the Fermi Gamma-ray Space Telescope. In September 2008, the Fermi telescope picked up a burst from around 12 billion light-years away, which had a spread of arrivals, depending on energy, over a time span of 20 minutes, which seemed consistent with the MAGIC data, though there were some concerns because the spread seemed too large for any likely effects of interacting with quantum space alone.

Later observations from the Fermi satellite put this interpretation into doubt, though. In 2009, Fermi observed a gamma-ray burst from around 7 billion light-years away, which showed no detectable lag. The feeling is that the earlier readings were probably a side effect of the process that produces gamma-ray bursts, and that any spread generated by quantized space is yet to be detected.

It is entirely possible that Fermi hasn't got the sensitivity to deal with such readings and we may have to wait for the next generation of device. Or it could be that the predicted interaction with the graininess of space simply doesn't happen, meaning this isn't a way to provide helpful information on how quantum gravity works. One possibility for making the distinction is the IceCube Neutrino Observatory at the South Pole.

This remarkable device, completed in April 2011, uses a square kilometer of ice as its detection medium, with detectors buried nearly 1.5 miles (2.5 kilometers) down in the ice looking for tiny flashes where incoming neutrinos (see below) collide with the ice above. It is just possible that IceCube could detect a quantum space-time shift in arriving neutrinos, as they can have even smaller associated wavelengths than gamma rays. It has also been suggested that if the LISA gravitational wave detector in space is ever built (see page 243), it could be converted (at considerable expense) to detect variations in the speed of light.

There are three aims for those who are searching the data from space for signs of quantum gravity. The ideal is to discover a clear way to distinguish between competing theories, though this is not possible at the moment because of the dearth of useful predictions from the theories. The other two aims sound similar. One is to demonstrate if gravity really is a quantum effect, and the other to determine whether space-time is quantized. It is entirely possible for gravity to come in a quantum form, but for space-time to be continuous.

One of these three has been achieved. For a long time it was taken on trust that gravity had to be quantized, but science requires evidence. An experiment with results published in 2002 did manage to demonstrate gravity acting in a quantum fashion, showing that reality matches expectations, and it was just a matter of getting theory to catch up. The experiment, at the French Institut Laue-Langevin made use of a quantum phenomenon called bound states.[15]

Quantum theory originally emerged from the realization that the electrons surrounding an atom could not exist with any old energy. Rather than having a continuous range of possible values, there were specific energy levels (called orbits) within which the electrons were constrained. If they gained or lost energy, rather than smoothly shifting from one level to another, the electrons made a quantum leap, jumping from one level to another by absorbing or emitting a photon, a quantum particle of light energy.

In the quantum gravity experiment, neutrons were used—uncharged particles from the atomic nucleus that aren't influenced by electromagnetic radiation, making sure it was gravity that was producing the effect. The neutrons were put in a chamber with a material at the bottom that would act as a mirror, so they would fall under the influence of gravity then bounce off the mirror surface.

If gravity were not a quantum force, the neutrons should accelerate smoothly down the chamber. But instead they were discovered to jump from one height to another—exactly the same behavior as electrons changing levels around an atom, though here the neutrons were assumed to be absorbing or emitting gravitons, the hypothetical carrier particles of gravity. All the evidence is that gravity is, indeed, quantized.

A more recent experiment at the same location has managed to push neutrons into different quantum gravity states.[16] A stream of neutrons is passed between two plates, a lower one acting as a mirror and an upper one that absorbs neutrons. The lower plate is vibrated at different frequencies, some of which match the energy jump required to push a neutron into a different state, changing the rate at which the neutrons that are not absorbed by the top plate emerge from the system.

At the time of writing the theoretical implications of this experiment were still being explored, but it should give the most detailed measurements of any variation between the expected force of gravity and the actual effect on this tiny scale. This may in the future provide an insight into which of the theories of quantum gravity best matches what actually happens.

One recent discovery that has emerged from the theoretical side is that many of the theories of quantum gravity predict a peculiar behavior for quantum space-time. It appears that on the very small scale, where quantum gravity is dominant, particles should behave as if they are existing in a two-dimensional universe—one of time, one of space.[17] Although this prediction came out of an obscure quantum gravity theory called causal dynamic triangulation, it proved to be a common prediction for string theory, loop quantum gravity, and others.

The prediction is based on an assumption that at the level of the atoms of quantum space (sometimes called quantum foam), there will be a distortion of space-time and one dimension is likely to dominate the others, leaving quantum gravity effectively acting in two dimensions (though the selected spatial dimension would not remain constant, but would keep changing directions). As yet this has not been confirmed but might enable a few of the more obscure theories to be eliminated.

Monitoring gamma-ray bursts may well give more clarity on whether or not space is quantized, but to get a clearer idea of the form of quantum gravity it would help a lot if we could see better into the early life of the universe. At the moment our principle view of the infancy of the universe is the cosmic microwave background radiation, which gives us a snapshot from around 370,000 years after the universe began. Before this point, the universe is thought to have been opaque to light because there was so much concentrated energy that all the matter was in plasma form, which means that looking through the universe would be a bit like trying to look through the Sun.

If we can penetrate further back in time—and if the quantum gravity theories can be tamed to make clearer predictions—then we may see some distinctions at that point, that help decide between the theories. One way to see closer to the big bang (or bounce) is by using neutrinos. These particles are produced in vast quantities in nuclear reactions. All the time we are bombarded by neutrinos from the Sun—around 50 trillion of them pass through your body each second.

The good news about neutrinos is that they are not stopped by plasma. They would enable us to see way back through the early life of the universe to within a second or so of its origin. But the bad news is that this very unstoppable character of the particles makes them extremely difficult to detect. Neutrinos were dreamed up to explain some missing energy in nuclear reactions back in 1930, but it wasn't until 1956 that one was detected by spotting the outcome of a very rare collision between a neutrino and another particle.

We do already have neutrino detectors. They are usually buried far underground in disused mines, using the screen of kilometers of rock to cut out any other particles that might be detected and cause confusion. The vast bulk of neutrinos pass through the detector as if it isn't there, but just a few interact with the atoms in the detection material (often cleaning fluid), creating a spray of particles and energy that can be picked up by sensitive detectors around the chamber.

A neutrino detector has already been used as a crude telescope to produce an image of the Sun. In a powerful demonstration of how little neutrinos care about intervening matter, the Sun was on the far side of the Earth when the image was made. But the result is like a picture from a very early computer game.

Composed of just a few pixels, it makes the Sun look as if it is made out of Lego. If there were ever to be a chance of using neutrinos to see back to the vicinity of the big bang, investigating quantum gravity when it was at its most obvious, we would need a whole new way of detecting and working with these slippery particles.

There is one mechanism, though, that can in principle penetrate even further into the past than neutrinos. We might get a better idea of just how the universe began and what form quantum gravity takes if we can ever detect gravitational waves. But despite millions of dollars being plowed into a whole range of experiments, as yet not a single gravitational wave has been detected.

CHAPTER TEN

PARTICLES AND WAVES IN THE ETHER

> *The tendency of modern physics is to resolve the whole material universe into waves, and nothing but waves. These waves are of two kinds: bottled-up waves, which we call matter, and unbottled waves, which we call radiation or light.*
>
> —*The Mysterious Universe* (1930)
>
> James Hopwood Jeans

Newton had been clear to say that he framed no hypothesis for how gravity managed to influence bodies at a distance. Einstein took away the problem by looking at gravity as less of force and more a change in the structure of space-time. This change can be regarded as radiating out from the body, each tiny portion of space-time influencing the next, so there is no need for action at a distance. Yet Einstein's general relativity tells us more.

Although there has been no success in bringing gravity into line with the other forces of nature, there are some similarities between the view of gravity provided by general relativity and the understanding of electromagnetism that arises from Maxwell's work. Faraday had shown experimentally the relationship between electricity and magnetism, and Maxwell explained theoretically that an oscillating electrical charge should produce electromagnetic waves—the source, for example, of radio.

Similarly, in 1918, Einstein produced solutions of his general relativity equations that predicted that a mass that undergoes oscillations should similarly produce gravitational waves—subtle variations in the gravitational field that propagate away from the source just as radio waves radiate from a transmitter. According to the equations, these ripples in space-time should travel at the speed of light.[1]

This makes sense. Take a pair of stars orbiting each other. As we have seen, such binaries are very common. As the stars move around, the joint gravitational pull felt at any particular point will keep changing, thanks to the altering position of the two bodies. The warping of nearby space will be quivering with the movements of the stars. These quivers should propagate through space-time as waves. If the waves arrived at a distant point instantly then we would receive information about the stars at faster than the speed of light—yet according to special relativity, information can't travel beyond the light speed barrier.

This means that in the (ridiculously unlikely) event that the Sun suddenly disappeared, not only would it take 8 minutes before we no longer saw it, but also the Earth would continue to orbit the nonexistent Sun for 8 minutes until the gravitational information arrived. Mercury and Venus would be hit sooner, but the Earth would sail on regardless. Thinking about the warped space-time caused by the Sun, we can imagine that this will take time to straighten out. It can't suddenly flip back out of its twisted form, but rather the straightening out of the warp will propagate through space-time at the speed of light.

If such gravitational waves exist and have any similarity to electromagnetic waves, we would expect to be able to treat them as particles. Electromagnetic waves are often better thought of as photons, and photons provide the vehicle by which electromagnetism interacts with matter. But was it possible to take a quantum approach to gravity waves? As we have seen, a full quantum theory of gravitation has so far eluded science, but this does not mean that it is impossible to find a version of Einstein's 1918 equations that allows for gravity waves.

Just over 40 years later, the British physicist Paul Dirac produced a version of the gravitational field equations that predicted there should be gravitons as the carriers of gravity, just as photons carry electromagnetism. When regarded as waves, just as with light, the energy of the quantum particle was proportional to the frequency—in fact it had the same relationship of energy: Planck's constant times the frequency.

There is some potential for confusion here. Photons have two roles in electromagnetism. They make up light, which can be considered as electromagnetic waves, but they are also present in a "virtual" form (because they are never directly observed) as the carriers of the electromagnetic force. When, for instance, a magnet attracts a piece of metal, we don't see light streaming between them, but virtual photons connect the two. Similarly, gravity waves are not an inherent feature of gravity, but rather a result of oscillating changes in gravity, just like emitted light, but gravitons would also be the carrier for all gravitation, not just gravity waves.

Gravity waves are transverse, like electromagnetic waves, meaning that the oscillation of the wave is side to side as the wave propagates, as distinct from a longitudinal wave like sound, where the wave vibrates in the same direction as the sound moves. This has significant implications. From the complexities of general relativity it emerges that such a wave will only be an oscillation in the warping of space, not of time. It also means that gravity waves should only be detectable in directions where the source appears to be moving side to side, rather than forward and backward.

Compared with light, observable gravitational waves are expected to be very low frequency. Visible light ranges between 4×10^{14} and 8×10^{14} hertz (this is the unit of frequency that used to be known as cycles per second, reflecting the number of complete cycles the wave of light goes through in a second). Gravitational waves depend on the speed of movement of the objects creating the waves. To be detectable these sources are likely to be stars or similar-sized objects, which limits the rate at which change can occur. Gravitational waves are expected to be received between 0.001 and 10,000 hertz.

Gravitational waves are more than an interesting quirk that emerges from general relativity; in principle they could be a very powerful tool in exploring the universe. Without light (in which I am including the whole spectrum from radio to X-ray and gamma ray, not just the familiar visible part), we would have no idea about the universe—what's in it, where it came from, or where it is going. The ability to detect light gives us a bridge to distant parts of the universe, and because that light takes time to arrive, it provides a visual time tunnel to examine the deep past.

However, light has its limitations. Visible light, for example, can't penetrate dust. By using a different energy of light—radio, perhaps—we can get around this problem. Even though using the different energy spectrums of light enables us to get around a lot of obstacles, each type of light has its own difficulties. As we have seen there are other particles that are less influenced by matter—neutrinos, for example—but we are limited to detecting these from bodies that emit them, like stars, and they are incredibly difficult to detect. Gravity waves offer us another possibility. You can't stop gravity by putting something in its path. With the exception of any antigravity devices we can dream up in the next chapter, gravity always gets through.

Unfortunately, gravity waves are extremely difficult to detect. Yet even before there was any possibility of detecting these waves, there was a form of confirmation that they existed. This seems a strange statement, but bear in mind it is a common enough situation when dealing with the universe. Many of the widely accepted components of our picture of the universe—the big bang, black holes, and dark matter to name but three—are not the result of direct observation, but instead are theoretical entities that are supported by indirect observations. And in the 1970s a discovery was made that provided the same kind of indirect evidence for gravity waves.

We have already met pulsars, the fast-rotating neutron stars that emit a pulsing beam of light, often in the radio frequencies. In 1974, American radio astronomers Joseph Taylor and Russell Hulse discovered a strange pulsar, later given the not-exactly-romantic name of PSR1913+16. They were using the massive fixed dish radio telescope at Aricebo in Puerto Rico and discovered that their new pulsar didn't just pulse, but the pulsation itself varied in frequency, as if the fast-spinning star was speeding up and slowing down.

This seemed very unlikely. It would take an inconceivable amount of energy to keep fiddling around with the rotation rate of a star like this, and particularly to do so repeatedly with a standard frequency. It seemed more likely that they were receiving a signal from a binary star system, of which the pulsar was one of the two stars, and it was the orbit of the pair of stars that caused the variation in the frequency of the pulses.

The other star has never been detected. It has to be small, because the frequency of the changes suggests a very tight orbit, sending the stars dancing around each other every 8 hours. However, the mass of the unseen star, implied by the orbit, is too small for it to be a black hole, so it appears to be a second neutron star that isn't emitting detectable radiation.

Such a system should be generating gravitational waves—it is exactly the sort of oscillating system that ought to be pumping them out, but there was no potential to detect such waves when the pulsar was discovered. However, gravitational waves carry energy with them. As the stellar pair continue to radiate, the result should be that these orbiting stars would slow down a little and the orbital distance would shrink, making the stars orbit each other faster.

This effect has been observed. The frequency of the variation in the pulse is going up, implying that the orbital period of the stars is reducing. And though we are talking very small measurements indeed, the rate of change matches that predicted by general relativity. This is not absolute confirmation of the existence of gravity waves, but the increase in frequency is useful encouragement that nature is behaving as it should if the gravity waves were there.

Of course, there could be another cause. This is always the problem with indirect observations. In the same way, the effects of the big bang, black holes, and dark matter can all be assigned to alternative sources. For example one alternative to dark matter is modified Newtonian dynamics, which suggests that bodies the size of galaxies don't move exactly the same way as "ordinary-sized" bodies and this explains the way they spin without the need for any extra mass. Similarly, the change in orbital period of the pulsar binary could have a cause other than gravitational waves—but the data does come up with the right values to support that theory.

This inability to be sure reflects a limitation of science, particularly when dealing with distant and indirect observations. Science can easily disprove a theory by producing data that are different from those that the theory predicts. But the scientific method can only ever offer *support* to a theory, rather than prove it. That same data can equally well support an alternative theory and the decision between them will then come down to which of these theories matches up less well to some other observation or experiment, or even to which theory is in vogue.

If and when we have technology that is sensitive enough to detect and make use of gravitational waves, there is a huge opportunity available—a chance to see beyond a barrier in time. As we have seen, gravitational waves would not find any limit in the plasma phase of the universe that prevents us from seeing optically beyond the 370,000-year mark. They should pass through regardless, enabling gravitational astronomers (if they ever exist) to see further and further back toward the very beginning, surpassing even the capabilities of neutrinos.

Using gravity in our study of the heavens is nothing new. Although the idea of detecting gravity waves is a relatively recent addition to the astronomers' toolkit, and as we will see is fraught with difficulty, they had realized earlier that conventional optical astronomy could be helped out by an effect of general relativity. Here the action of gravity is not the carrier of information from distant stars, but is rather acting as a part of a celestial optical instrument.

We see remote objects through a telescope by a process of collecting light across a wide area and focusing it to get an image. Instead of merely seeing the result of photons traveling in a single straight line from the object to the eye, a telescope picks up different beams of light that hit various parts of the collecting device, pulling those beams together and focusing them at a point, strengthening the image many times over.

Such focusing requires a device that bends light. Galileo's telescope did this using a lens. Light bends when it passes from air to glass (and vice versa) in the process known as refraction. Because a lens is curved (literally shaped like a lentil), light hitting different parts of the lens is bent by different angles, pulling the rays of light together to form an image.

From early on, astronomers realized there was a problem with using lenses this way. Different colors of light are bent by different amounts as they pass from air to glass and glass to air. This is why a prism separates white light into a rainbow of colors. When white light passes through a lens it similarly tends to produce rainbow fringes, distorting the image in a process known as chromatic aberration.

Although later astronomers would develop combinations of lenses to minimize these colored distortions, the alternative approach, championed by Newton and still used on all large optical telescopes is to replace the main light-collecting lens with a mirror. This focuses the light rays, but bends all colors equally. (Mirrors also have the advantage of having a shorter focal length for a large light collector. The biggest refracting telescopes, using lenses, had to be ridiculously long.)

A similar approach to a reflecting telescope is taken in the design of radio telescopes. Here, rather than a metal or glass and metal mirror, the collecting device is an immense dish that is used to focus the radio waves onto a receiver, usually suspended high above the dish at its focal point.

Now think of what happened in the observations made during the total eclipse of 1919 that helped established general relativity. The light from the observed stars was bent inward toward the Sun. All we have to do is think much bigger and better to provide a whole new focusing mechanism that can pick up very distant light and focus it for us. Imagine we want to look at a very distant galaxy and it happens to be behind another galaxy. This sounds like a problem, but thanks to general relativity it can be a real advantage.

Imagine the light from a distant galaxy heading out past the closer galaxy in a direction that would miss the Earth. The vast gravitational pull of the closer galaxy will bend the light from the distant galaxy inward. Get the position just right and the closer galaxy will act like a lens, producing an image of a distant galaxy that is otherwise invisible. Many of the different light rays that pass around one side of the nearby galaxy will be brought into focus, as if it were the objective lens of a giant telescope.

That's the good side of using gravity as a lens—but it has a negative aspect, too. These gravitational lenses can result in multiple images of the same object in the sky, causing potential confusion for astronomers. In fact, gravitational lensing was first demonstrated in action in 1979 when the same quasar was discovered in two locations on a radio frequency scan. What is happening here is that the intermediate object is massive enough to bend the light around both sides of it. Light from the left side will tend to produce an image shifted to the right and vice versa.

The most dramatic effect is to produce halos, known as Einstein rings, where the image of a single object is distorted into a complete circle, or an arc, by the warping effect of a strong gravitational field. Here, the light from a distant point of light is being bent around all sides of the nearer focusing mass. The light beams aren't bent far enough to bring everything together at a single point, but they do pass through enough of a warp that they are all visible, producing a ring around the nearer object.

But gravitational lenses merely enhance (or distort) the use of light. Gravitational waves are quite different. In principle they could replace light in a telescope altogether. And whether we are trying to penetrate the secrets of the origins of the universe or to see through an otherwise impenetrable cloud of dust or supermassive black hole, there is little point recognizing that gravitational waves can go where no light can if it isn't possible to detect these ripples in space-time. Unfortunately, finding them is not a trivial task.

A gravity wave detector has to spot fluctuations in gravitational force that amount to perhaps 1/1000000000000000000000 of the typical level.[2] At the same time—and this is an even bigger problem—the gravity telescope has to screen out all the fluctuations in gravitational pull that aren't being caused by gravity waves. This level of accuracy is comparable with the difference in mass between a person and the Earth. A car driving past would produce a significantly larger fluctuation in a gravitational reading than a typical gravity wave.

There is also another problem facing designers of gravity wave detectors. When we use an optical telescope we know what we are looking at, because the device is pointed at a particular part of the skies. All light-based telescopes can be aimed at a potential source. But many gravity wave detectors would pick up any waves that arrived in a particular plane, making them susceptible to billions of sources simultaneously. This means that an effective gravity telescope, rather than just attempt to prove the existence of gravity waves, would have to make simultaneous observations in multiple orientations to home in on a source.

To make matters worse, as gravity is not blocked by matter, any working detector would pick up sources on the other side of the Earth just as easily as sources on the side where the detector is located. Being absolutely certain of linking up a gravity wave detection with a specific source in space would be a nightmare.

What is surprising in a way is that, knowing just how sensitive a gravity wave telescope would have to be, there have been attempts at building such detectors since the 1960s. That variation produced by the typical gravity wave from space will produce a difference of 1 in a measurement of 10^{21}, 1 with 21 zeros after it. When the first gravity wave detector was assembled it was a factor of 10 million out. It could only detect a variation of 1 in 10^{14}.

It might seem that the American physicist Joseph Weber,

who devised this first detector at the University of Maryland was wasting time (and quite possibly taxpayers' money), but there were two arguments for going ahead with the experiment. It was possible that the estimates on the limits of gravitational waves were wrong, and they could get a surprise—this has happened in the past with other experiments. And lessons would be learned from the first device that would make it easier to develop better gravity telescopes in the future.

The original detectors like Weber's were just heavy metal bars, circular in cross section, which were isolated as much as possible from the vibrations and gravitational influences of the world around them. The hope was that, divorced from their surroundings, the bars would receive the incoming gravitational waves and begin to vibrate themselves, resonating in a similar fashion to the way a piano string will start to hum if its note is played or sung nearby. These two-ton cylinders of aluminum alloy never found anything.

Weber was not put off. He built a series of detectors, pushing the sensitivity up to around 1 in 10^{16}. This was still highly unlikely to produce any results, yet Weber reported picking up a wave simultaneously on two separate detectors, which should have reduced the chances of any local cause. By placing the detectors at two widely separated locations, Weber should only be picking up vibrations that came from a considerable distance away (though this did not rule out some kind of resonance in the Earth itself).[3]

At least, this was the deduction that could be made if the

detectors were perfect and if it were possible to eliminate all stray vibration. But the reality is that glitches were always going to slip through. These bars would be subject to movements that had no significance as far as gravity waves were concerned. In communications terms, such readings would be noise rather than signal. And some of these random signals would coincide on the two detectors by chance. But Weber had thought of this and used a statistical technique to reduce the chances of noise being mistaken for a signal.

Each of the two detectors will occasionally detect a signal. These are only of interest if they occur simultaneously—anything triggering only one detector is likely to be local. With a dual detection, there's a chance we are dealing with a real gravitational wave. Although it would be ideal if there were some regular pulse being measured as ought to be detected from a continuously produced wave, say from a binary pair of stars, realistically, operating so close to the limits of detection, it is more likely that occasional spikes would be all the detectors could manage.

The technique Weber used was to compare outputs from the two detectors, but to shift one of those outputs gradually in time compared to the other. Once one output had been shifted, the detected peaks that were simultaneous would no longer be alongside each other. He carried on shifting the time base of one signal until another pair of peaks lined up. Because of the time shift, he knew that these two peaks weren't from the same signal, but were a random coincidence.

By repeating this process again and again, Weber hoped to

build up a picture of the chances of two peaks occurring randomly together. He was then able to compare the actual rate of matching between the two signals with the random generation of coincidences that came from shifting one signal in time. He could see if there were more coincidences of signals than would be expected from noise alone.

This isn't a definitive way of identifying positive signals, but it does set a threshold below which results ought to be ignored. And as far as Weber was concerned, the result of this coincidence testing was positive. His peaks that turned up on both detectors, he decided, were unlikely to be just noise. They looked like the signature of an incoming gravity wave.

When a scientist produces an unexpected result that could have major implications in a particular field, the usual process is for the experiment to be replicated by different scientists in other labs around the world. This reduces the chance of error on the part of the experimenter or the experimental design, and helps remove the possibilities of local interference. Traditionally, until such replication has confirmed the result, it is not made widely known. The alternative is to generate bad publicity for science.

This is part of the reason that the scientific community came down so heavily on the experimenters Stanley Pons and Martin Fleischmann when they publicly announced that they had produced cold nuclear fusion in a test tube before there was any chance for others to replicate the experiment. Going public in this way was considered bad form—the sort of thing that pseudoscientists do. When no one else could produce the same results, Pons and Fleischmann were vilified.

The need to replicate has interesting implications that are rarely mentioned for "ultimate experiments" like the Large Hadron Collider in CERN. There is no other piece of equipment in the world that can produce the same results as the LHC. So should it provide breakthrough information, technically that data will not have been verified. Of course there is a small army of scientists working on it and multiple runs of the equipment are used to see if the same results come out every time. It's not as if a shock result will be the one-off work of a single oddball team. But even so, it is slightly unnerving that there is no opportunity for testing of results by replication.

Weber's gravitational telescope was capable of replication, and with such dramatic results coming through it was inevitable that other groups would look out for these remarkable gravity waves. Nothing was detected. Of course there were vibrations, but all of them could be explained away as a result of local causes or coincidence. The gravitational waves that Weber thought he had detected refused to appear for anyone else. While it is impossible to be certain, it seems likely that they were spurious, caused by some failure in the equipment or a misleading interpretation of the data.

With a certain stubbornness, this kind of detector continued being built for decades after Weber's original. Over time they became better isolated from the environment, including introducing supercooling to the bars to reduce any thermal vibration, as the natural movements of the atoms within the bar will be more sluggish at low temperatures. Such detectors have now just about reached the level where they may be able to detect the fringes of the predicted range for gravity waves, though they are yet to produce a single unchallenged positive result.

It must have been a frustrating 40 years for the scientists involved. Imagine building a device at some point in the 1980s. You know that your detector will be better than Weber's, perhaps by a factor of one hundred or more. Yet even so, if accepted theory is correct, you will never detect anything. It takes a special kind of scientist to persevere under such conditions.

Although instruments measuring resonance in solid bars have continued to be made, because they are relatively cheap and easy to construct, most recent efforts to produce a gravity wave detector have gone into a different style of equipment called an interferometer. Interferometers make use of interference patterns in light to measure tiny shifts in distance. A typical setup might send two beams of light at right angles down tunnels in an L-shaped device. After repeatedly traversing the legs of the L, the beams are brought together using mirrors.

If you think of the beams as waves, when they are brought together, any waves that are in step will reinforce each other and you will get a more intense beam. If the waves are totally out of step, so one is down while the other is up, they will cancel out, reducing the beam strength. This is interference. But move one of the mirrors by a part of the wavelength of the light and one beam will move against the other, changing the interference pattern. This way, tiny shifts in the mirrors, much smaller than the wavelength of light can be measured.

The wavelength of visible light is around 1 millionth of a meter (about 0.0004 inches), with higher energy light even shorter in wavelength. Combine this with a very long measuring device and remarkably small shifts can be detected. The hope is that a gravity wave would disturb the mirrors, causing them to move out of position, a measurement detected by the effect on the light beam, and quantified by the force required to bring the mirror back into position, which would be used to measure the strength of the gravity wave.

It might seem obvious that a gravity wave would move an object with mass, but it took some argument before this could be agreed on—the scientists involved are working with very theoretical constructs. The problem is, a gravity wave is a variation in the warp in space-time caused by gravitation. As such it won't only influence the mirrors, but everything around them. There were some concerns that gravity waves would be invisible because anything used to make a measurement would change just as much as the thing being measured, but theorists assure us this isn't the case.

Monitoring changes in an interferometer in this way is the premise of LIGO (Laser Interferometer Gravitational-Wave Observatory). LIGO is not a single device, but a pair of instruments that are connected together by computer to make them act like a vast detector. (LIGO is not the only gravitational interferometer experiment worldwide, but it is the most sensitive, so I'll stick with it as my example.)

Both interferometers consist of a pair of tubes, each around 2.5 miles (4 kilometers) long, in an L shape. These tubes are evacuated of air, then a laser is shot up and down the tubes 75 times to amplify the result, before the beams are brought together to form an interference pattern. One device is in Hanford, Washington, the other in Livingston, Lousiana, around 1,800 miles (3,000 kilometers) apart. (There is also a second, smaller device at Hanford to check the results.)

These are very sensitive devices. Just as with the bar resonance detectors, it has been necessary to eliminate any local waves that could be misinterpreted—the interferometers can easily pick up waves in the sea, rolling onto the beach, for example. LIGO went operational in 2002, 10 years after the project was begun. At the time of writing it has still to produce any useful positive results.

It should be stressed, however, that negative results are also useful, as they can disprove a theory, but as yet, the detectors are not sensitive enough for the negative results to rule out gravity waves. There may also already have been gravitational waves detected by LIGO as several years of data is still under analysis. But it is touch and go whether the current version of LIGO will ever produce positive results.

The best the gravity wave researchers have managed so far is a gradual advance of upper limits. This is something we hear more about in the news from particle physics, with searches for missing particles like the Higgs boson. Every now and then it will be announced that it has now been shown that the Higgs boson does not exist within a particular energy range—the search is narrowed. Similarly, the negative results from LIGO have put upper limits on the strength of gravity waves in particular frequency bands. We know that they can't be above a certain strength, because they would have been detected. It's not a lot, but it is a crumb of a result for all the work that has gone on.

There is a small hope that gravity waves could be detected by adding extra detectors of similar sensitivity to the current setup. As well as LIGO and the smaller European VIRGO and GEO600 detectors, at the time of writing there are three planned new detectors in Japan, Australia, and India. According to the journal *Classical and Quantum Gravity,* just one new detector should more than double the detection capability, while bringing all three online could result in 10 times the capability to spot gravity waves, but even at these levels it would be entirely possible that no gravity waves would be detected, and funding could be cut before the detectors are ever built, as these new projects were announced at a time of world recession.[4]

The ever-patient gravity researchers also hope to rejig the existing equipment to produce an Advanced LIGO, installing more sensitive detectors at existing locations. Planned to come on stream in 2014, Advanced LIGO should have a better hope of producing a result, giving each detector around 10 times the sensitivity of the current devices, but to be certain of testing out the existence of gravity waves one way or another, the long-term hope is to move into space.

As the Hubble space telescope has proved, being out in space is a boon for astronomers. It removes your telescope from all the distortions caused by air movement and pollution. The instruments are kept away from weather and away from the vibrations that are always passing through the Earth, whether from natural tremors or human activity. In scientific terms, the benefits of all the manned missions are negligible (though admittedly a Shuttle mission did fix a fault on the Hubble). Almost all the scientific knowledge we have gained in space is from unmanned probes.

Gravity wave enthusiasts would like to follow Hubble and take a gravitational telescope into space. There are two great advantages for the space equivalent of LIGO, called LISA (Laser Interferometer Space Antenna). First, just like Hubble, it gets the telescope away from all the local noise. While a gravitational telescope isn't bothered by clouds, it can be disrupted by storms, but even more so by all the tiny vibrations that are constantly being emitted around us. Being in space produces a much greater degree of calm.

Out in the solar system, things aren't entirely without disruption, of course. There are solar winds, streams of particles flow from the Sun, cosmic rays, and other high-speed projectiles in space that would cause disruption that needs to be taken into account. But it is still a much more isolated and controlled environment.

The second benefit of shifting the detector to space is one of scale. LIGO's interferometers use 4-kilometer (2.5-mile) beams. LISA is planned to have a triangular arrangement with beams 5 million kilometers (3.1 million miles) long. LISA would work at a different frequency to LIGO, so apart from being more sensitive, it would also expand the range of frequencies of gravitational wave that could be spotted.

At the time of writing, LISA is currently still at the proposal stage. Back in 2005 it was expected to be live by now, but realistically we couldn't now look at a launch before 2025. To make matters worse, LISA was originally a joint NASA/ESA project, but NASA is likely to withdraw due to funding limitations and ESA will be reviewing the project's future during 2012. The project is very likely to be dropped.

We really ought to admire the scientists who manage to sell these experiments to the politicians in the first place. It's not that gravity wave detection is unimportant. It could eventually give us invaluable information about the nature of gravity and a way of finding out more about the origins of the universe. But to have to go to a funding meeting and defend spending millions of dollars on detectors to follow up on previous experiments that have detected nothing whatsoever for more than 40 years takes considerable chutzpa.

It is also the case that the constant effort of working on a project for year after year without producing anything must have an abrasive effect on the human psyche. It is fascinating to read the account of the work on gravity waves by Harry Collins, a sociologist who has studied the work of those engaged in the search. Collins feels obliged to defend them, suggesting that there is somehow "easy science" like Newtonian physics, quantum theory, and general relativity, and "hard science" like the search for gravity waves.[5]

Those who have attempted to study quantum theory or relativity may doubt Collins's assertion of the easiness of these theories, but he is not referring to the difficulty of understanding the math, but rather the way these fields are capable of accessing exact results (okay, quantum theory is all about probability, but it is *exact* probability). He contrasts this with something like weather forecasting, which is not capable of being approached in this way, but must always be an approximation that fast runs away from accurate prediction.

This sounds like a profound misunderstanding of science by a sociologist—weather forecasting obeys exact values as much as any other science, it is simply that there are far too many variables to handle the exact values. To use this as a defense of the inability to find gravity waves seems more than a little unfortunate. However, he does have a point that the limitations of the technology mean that scientists working on gravity wave detection have to take a more subjective approach than is usual (or desirable) in science.

What should happen in a worthwhile experiment is that a measurement is taken (give or take a margin of error), knowing that the measurement is relevant to the phenomenon being studied. By contrast, in current gravity wave experiments there are multiple problems in the way of being sure that what is being detected is what the scientists are looking for. As we have already seen, there is a statistical approach taken to deducing whether matching signals on the two detectors are the result of a genuine gravity wave or coincidence. A variant on this approach is still used with today's interferometer detectors.

While there is nothing wrong with using statistics—scientists have to do it all the time—there is the risk of a one in a million occurrence actually happening and causing a false measurement to be taken as real. In most experiments this isn't a problem as the process is repeated over and over again, so such false measures are likely to be identified as invalid data. But where the equipment is operating at the very edge of detection, where events are very rare, it is much more likely that the statistical approach will over-value a false signal.

The subjectivity also applies to the filtering out of obvious causes. If, for example, there is a major earthquake, the vibrations are felt all over the Earth. Both detectors in a gravity wave observatory will pick it up, and if the locations are equidistant from the source of the tremor, they will pick it up simultaneously. The scientists monitoring the devices are expected to eliminate such a vibration as a matter of course because they are both monitoring using traditional seismometers and keeping on top of the information channels that cover such occurrences.

So the teams will flag up a particular vibration as not being useful data. But here's where the subjectivity comes into play. How far should they go in flagging dubious data? If they were to eliminate every data point that had the faintest possible source (someone is using a jackhammer 100 miles away, for instance), they would probably never accept any measurements.

As a result, the scientists have to set arbitrary limits of exclusion. Some data they will use, some they will ignore. This arbitrariness leaves them open to accusations of subjectivity and cherry-picking the data to produce the result they were hoping for, consciously or unconsciously.

To make matters more interesting, the LIGO teams have also had to face the possibility of a blind injection.[6] This is a mechanism to test the effectiveness of their systems at eliminating noise and picking out signals. At some point in time, fake signals may (or may not) be injected into the system. Unless the teams could identify such an injection as a true signal rather than noise (even though it is fake), their statistical manipulation of the data has to be considered flawed.

The use of this technique seems to have had mixed effects on the teams working on LIGO. It probably did increase their focus on detecting results, where previously, given the inevitable assumption that they may never find anything, they might have been more focused on discounting results. But at the same time, the possibility that the most interesting data they have could be fake would have reduced the excitement at finding an encouraging result. However good a reading may be, there is always the possibility that it is a blind injection at work.

Famously, when the SETI (search for intelligent life in the universe) program was at its height, there was a dramatic 72-second signal picked up by a radio telescope at Ohio State University. The researcher working on the data, Jerry Ehman, checking the printout from the scan, circled the section relating to the peak and labeled it with one of the most famous marginalia in science. It is now referred to as the "Wow! signal."

The downside of the threat of a blind injection is that it seems to have removed any potential for a "Wow!" moment in gravity wave detection. As one of the scientist working in the area commented about the blind injection, "All your enthusiasm gets sucked away . . . It's messing with your head . . ."[7] The danger is not just a lack of joie-de-vivre, but a real reduction in the sense of urgency and importance given to an interesting signal as it could always be the gravity wave equivalent of a mystery shopper at work.

In the event, after months of work analyzing data and in particular after spending a huge effort on one event that looked promising as an actual gravity wave signal, the truth was revealed about the blind injection. Not only was the promising signal that they had spent so much effort analyzing a fake event, but there was a second blind injection no one had spotted. So the only significant-looking piece of data was artificial, and an equally interesting bit of manufactured data was missed entirely by the teams.

This doesn't make the efforts of the gravity wave search totally worthless. However it does emphasize just how little real data there is when the only major point of interest was a blind injection. And it also stresses how close the signals are to the edge of detection when what should have been a significant signal was overlooked.

Arguably this arbitrariness suggests that it was pointless spending millions upon millions of dollars on LIGO and they should have gone straight to LISA, isolated from most of the vibrational interference of nature and human activity. It is entirely possible to argue that the gravitational wave scientists have acted a bit like children faced with a difficult choice.

If you ask a child to choose between an item they can have today and one they have to wait a month for, saving up a significant part of their allowance for it, they will usually go for the "today" option. This is even the case if you point out that the immediate gratification option won't really work. But if they wait and save up their money, they can get something a lot better that will do what they want it to do.

There seems a suspicious parallel in the decision of the science teams who have supported Earth-based gravity wave detection. They have opted for the immediate dime-store solution (if you can call a project that has cost hundreds of millions of dollars "dime store") rather than save up and wait for the quality solution. And now, with financial constraints tighter than they have been for many a year, there may be a feeling of it being one spend too many.

If this happens it would be a shame. LISA really would help sort out the gravity wave argument in a way that ground-based systems never can. But it is understandably hard to get politicians who control funds excited about such an apparently abstruse piece of research. The same, however, can't be said about the search for an antigravity device. If antigravity could be demonstrated, then there would be plenty of government excitement (and money) following in its wake.

CHAPTER ELEVEN

CAVORITE RETURNS

> *Now all known substances are "transparent" to gravitation. You can use screens of various sorts to cut off the light or heat, or electrical influence of the sun, or the warmth of the earth from anything; you can screen things by sheets of metal from Marconi's rays, but nothing will cut off the gravitational attraction of the sun or the gravitational attraction of the earth. Yet why there should be nothing is hard to say.*
>
> —*The First Men in the Moon* (1901)
>
> Herbert George Wells

Pioneering science-fiction writer H. G. Wells gave us the first example of traveling through time by using a machine that exploited the concept that time is a fourth dimension. For many people, time travel is simply that—fiction. And it's certainly true that Wells's time machine works on fictional principles. But this long-time staple of science fiction is more possible than it seems. There is nothing in the laws of physics that prevents us traveling in time.

More than that, relativity (both special and general) provide actual mechanisms for time travel. They may be tricky in engineering terms to make happen on a useful scale, but time travel devices are perfectly feasible. In fact, thanks to relativity, every time you move, you travel into the future—and whenever you are in a gravitational field, you move slower through time than someone in empty space.

But this wasn't Wells's only dalliance with modifying space-time. He also originated another staple of science fiction in his book, *The First Men in the Moon.* This is an altogether lighter story than his dystopian time-travel story. Literally lighter. The hero travels into space in a ship designed by the inventor Cavor. The outside of the ship has blinds coated in a substance called cavorite, which blocks the attractive force of gravity. Close the blinds and your ship floats off into space; open them and it is attracted to the closest massive body.[1]

The whole idea of antigravity is a science-fiction favorite and a frequent topic of discussion for UFO fans and conspiracy theorists alike. If there were to be a substance like cavorite, it would have to act as a shield against gravity. This is all very well as a concept until you think about what it is that you are trying to do. If, as general relativity shows, gravity is the product of a twist in space and time, how can you shield against such a warp by putting something in the way? It's like trying to stop someone twisting your arm by closing your eyes.

Before we explore the real possibilities of antigravity, though, we ought to consider its more mundane twin, artificial gravity. There isn't much demand for this on Earth. Gravity is just . . . there. No need to make it. But it becomes more of an issue in space. As we've seen (page 7), humans and other living inhabitants of Earth don't thrive if kept in low gravity for a long period. Any long space journey is likely to require some form of artificial gravity.

Do-it-yourself gravitation certainly features in the majority of science fiction. On a TV show like *Star Trek,* artificial gravity is primarily there to avoid having to spend dollars on expensive special effects of floating around in microgravity. However it would make sense to have some way to keep things firmly on the "ground" both for practical and medical reasons. The writers often resort to fictional gravity generators that appear to manipulate space-time in some fashion to create gravity, but a real long-distance spaceship is more likely to resort to the equivalence principle.

There are two basic ways to provide gravity through equivalence. The most straightforward is through constant acceleration. If your spacecraft, once out of the Earth's gravitational clutches, constantly accelerates at 9.81 meters (32 feet) per second per second, then everyone onboard will feel 1 g of acceleration toward the rear of the ship. It will be exactly as if the ship were on the ground. What could be simpler than that?

At least, it seems simple. However, taking this approach comes up against the same problem that is faced by anyone trying to build a time machine to travel far into the future using special relativity. According to special relativity, the closer you get to the speed of light, the slower time passes for you, compared with clocks on the Earth that you are speeding away from. So taking a journey on a spaceship at near the speed of light will also take you on a jump into the future.

When the numbers are plugged in, though, it requires a phenomenal amount of energy to get a spacecraft up to near the speed of light, so any time-machine builder is faced with the difficulty of getting hold of sufficient energy to power the ship. It might seem that a gentle acceleration of 1 g is an entirely different prospect, but if it has to be applied for years at a time it would still require a huge amount of energy.

Our current space probes are blasted up to their full speed in minutes, but then drift along with no acceleration. To keep the acceleration going constantly would strain any practical resources to the limit. Even using our most compact form of energy, a drive based on combining matter and antimatter, it would only be practical to carry enough fuel for about a year at a constant 1 g thrust.

There are two problems with getting much further. One is that the mass of fuel that has to be carried starts to require a significant amount of fuel just to move it. But more significantly, as you continue to accelerate you would get closer to light speed. As this happens the mass of the ship increases and the amount of energy required to move it shoots up. It is not practical to keep up a 1 g acceleration for years at a time.

The alternative approach to artificial gravity is to spin the ship. If the ship is rotating, the natural tendency of people rotating with the ship is to shoot off sideways. They can't do this because they hit the ship's hull, so they are stuck to the hull by a kind of artificial gravity. It's often called centrifugal force, though many physicists will tell you this is a misleading term. The force being applied to you is not outward, but inward (centripetal force, as Newton called it). You are moving outward, but the hull applies a centripetal force to stop you moving.

Many of us have experienced this effect in action on fairground rides. The clearest example is a cylindrical machine where the riders stand in a circle with their backs against the curved wall. The cylinder starts to rotate at high speed and the floor drops away. Rather than fall with the floor, the riders are stuck to the wall of the cylinder by a kind of artificial gravity.

This is equivalence at work again. Acceleration is not just a change in speed, it's a change in velocity. Although we use the terms interchangeably in English, in physics speed and velocity are entirely different. Speed is a scalar—it is just a number. Velocity is a vector—it combines a speed and the direction of that speed. So acceleration is involved whenever something is rotated, because the direction is changing even if the speed isn't.

Although there is acceleration involved, we don't have the same energy problem as a ship that is constantly accelerating in a straight line. Once a ship is spinning at the right speed, it will keep spinning unless something opposes it, thanks to the conservation of angular momentum. Even though there is a force, keeping the people pressed against the hull, there is no energy expended. This is because the energy involved is the force times the distance moved in the direction of the force. But the rotating movement is at right angles to the force; there is no distance moved, so no energy is consumed.

This means a spinning ship can generate a useful form of artificial gravity. But it would have to be a large ship. We feel dizzy on a fairground ride because the sensors in our inner ears that allow us to balance and sense gravity are feeling a constant, rapid change of direction. If the ship were very large, then the rate of change of direction will be much smaller and there will be less discomfort. Spinning a ship could be an answer for artificial gravity, but we would be looking at very different designs to current space probes.

While creating gravity has obvious solutions, removing it remains both a hugely desirable dream and a practical nightmare. Even if general relativity's explanation of gravitation were not an issue, a cavorite approach to antigravity has the problem that it makes it possible to produce a perpetual motion machine, defying the first law of thermodynamics, which means that you can create energy from nowhere.

Imagine a simple machine with a set of paddles like a waterwheel. Coat each paddle on one side with cavorite. These sides will rise as the paddles will be shielded from the Earth's attraction, then, when they travel over the top of the wheel, will fall because the unshielded side is once more exposed to the Earth's gravity. There is no reason why such a wheel would not continue rotating forever, which any understanding of the conservation of energy makes highly unlikely. It would be possible to generate energy indefinitely, effectively from nothing.

It is easy to get carried away by the parallels between gravity and electromagnetism. We can, indeed, make a barrier that is proof against electromagnetism, a shield produced by effects within the atomic structure of the shielding matter. For example, a Faraday cage—a simple metal cage—will prevent electrical charges getting through to the inside. This happens because of the different charges on particles. The free electrons in the metal move under the influence of the applied electrical field and cause a new electrical field that stops the invading field from penetrating.

Unfortunately for those who want to build a gravity shield, gravity is not polarized in the same way. There aren't particles with both positive and negative mass that can be used to set up the gravitational equivalent of a Faraday cage to protect the contents from the influence of gravity. Yet this has not stopped a massive amount of effort, both from amateurs and from defense establishments, being put into exploring the possibilities of anti-gravity.

There are ways, of course, of reducing the gravitational pull felt by any particular object. The simplest approach is to apply an opposing force. Think of a helium balloon floating into the sky. What does it mean that the balloon floats upward? We know that force of gravity is proportional to the balloon's mass, so does the fact that it moves up not down mean that the balloon has negative mass? No it doesn't.

Mass is a measure of the amount of stuff in an object. If we break down the atoms to their component parts, you can consider mass to be the number of neutrons, protons, and electrons that go to make up the object (with a bit of twist thrown in for relativistic effects if anything is moving quickly). The rubber or metalized plastic of the balloon and the helium gas inside it are all made up of atoms, so the balloon has a positive mass from all those atoms.

Some people will tell you that the balloon's mass is positive, but its weight is negative. This, too, is wrong. Weight is just a measure of the force of gravity. As we've seen earlier, it is purely arbitrary that we measure weight in the same units as mass. Really weight should be measured in units of force (which would be newtons in the metric system). So when we ask how much a new baby weighs, the answer really should be something like 35 newtons. Weight should always be mass times the local acceleration due to gravity—it can't suddenly become negative.

What is happening with the balloon is that there is more than one force at work. If I put a heavy object on a scale, then pull upward on the object, the reading on the scale goes down. It doesn't mean that the object has suddenly started to weigh less (or for that matter that the force of gravity is decreasing). I am applying a force upward with my hand that counters gravity's pull.

Similarly a helium balloon feels a force upward because of its buoyancy. Just as a floating object is pushed up by the water it displaces, a balloon is pushed up by the displaced air. If the balloon is lighter than the air it displaces, that upward force is bigger than the downward pull of gravity and it floats. Gravity and weight are not changing, there is just a second force involved.

One extreme way to cancel out the pull of gravity from one body is with a matching gravitational force. If I place another Earth of the same mass immediately above your head you would no longer feel the pull of gravity, but would float between the two. (Not for long, admittedly, as you would be crushed when the two Earths collided under their respective gravitational pulls, but you would briefly experience a kind of antigravity.) However, this is hardly practical for building an antigravity device.

Similarly it is possible to use the equivalence principle to counter gravity with acceleration, because the two are indistinguishable. This is what happens when passengers fly on a "vomit comet" and experience weightlessness when the plane is in free fall. And it is equally possible to use the other easily accessible force of nature, electromagnetism, to counter gravity.

If you take a plastic comb and rub it on your hair, then bring it near a pile of small pieces of paper, each the size of a fingernail or smaller, the paper fragments will defy gravity and leap up to sit on the comb. When you rubbed the comb on your hair it picked up electrons, leaving the plastic negatively charged (and your hair positively charged).

A few moments later, you bring the comb near to the paper fragments. Now the negative charges on the comb push electrons away from the surface of the paper. The result is that the paper surface becomes positively charged and the paper is attracted to the comb by the electromagnetic force between the negatively charged comb and the positively charged side of the paper.

Here is where the disparity in strength between gravity and

electromagnetism becomes very obvious. Physicists tell us that gravity is a very weak force, but it hardly seems to be feeble when it keeps us on the planet, and maintains the Earth in orbit around the Sun. Yet the tiny electrical charge on your comb is enough to overcome the entire gravitational pull of the massive Earth. Electromagnetism wins hands down.

Although it is strange that gravity is so weak, it is just as well. If gravity had been much stronger it would have overcome the expansion of the early universe and everything it contained would have collapsed back into oblivion without ever getting far enough to form stars or planets. Even now, when the expansion of the universe is speeding up, gravity can locally overcome that expansion. The Andromeda galaxy, our nearest large neighbor, is moving toward us rather than away, like the other galaxies. But with a bigger gravitational force everything in the universe would have collapsed back out of existence.

The trick with the comb and paper can be duplicated on a much larger scale. Overcoming gravity with electromagnetism is not just a matter of lifting scraps of paper (or even scrap iron in a junk yard with a large magnet). In a complex form, it is used in maglev (magnetic levitation) trains. Here a strong magnetic field is set up so that there is repulsion between the train and the track. This lifts the train above the track, leaving it floating without direct physical contact with the ground. By sequencing pulses in a series of electromagnets it is also possible to use the electromagnetic field to propel the train along, though this part of the system is not an antigravity effect.

The first commercial maglev train was run in the UK as a link between the city of Birmingham and its airport, in use between 1984 and 1995. The system was withdrawn because of unreliability—with more complex technology than a traditional train, this is a technology that is still in its infancy. A handful of other lines have been opened around the world, and over time the technology is becoming more practical. Maglev has the advantage of removing the friction from contact with the track, and having no moving parts, but because it requires complete replacement of railroad infrastructure it could be many years before it becomes widespread.

Although maglev is an example of using electromagnetism as an antigravity force it is inevitably limited because it requires both a vehicle and a track to be involved in the magnetic system. It won't work between a plane and the ground, say, because the ground isn't a good enough conductor and a plane couldn't produce a strong enough field to get it high into the air (maglev trains ride very close to the track).

However there have been examples of magnetic levitation where the ground-based part of the system produces the field and the item to be levitated is not metallic. The most famous examples feature levitating frogs. In a frequently repeated experiment, these hapless amphibians have been suspended, floating in the air, above a powerful magnetic field.

Unlike a metal object, a frog doesn't have lots of free-floating electrons to help provide the repulsion, but it is made up of atoms, and atoms are capable of acting as tiny magnets in their own right.

Hit it with a powerful enough field and those little magnets will line up and repel the initial field. Frogs tend to be used because they are relatively light and have a high-water content—the water molecule is particularly effective at producing this "diamagnetic" field—one that is set up within an object in opposition to an external field.

In principle the same process could be used on a human being (we also have a pretty high water content), but it would require a much stronger magnet to support the extra weight. It would also be tricky because the electromagnets needed to produce the magnetism—usually a Bitter electromagnet, which uses metal plates rather than coils of wire—would have to be extremely large to fit a human inside the helix of plates that produce the field.

Although the frogs appear to suffer no side effects from their experience, there would also be some concerns about the risk of putting human beings into such a strong magnetic field. We know that powerful magnets can influence the brain. Although a static magnetic field should not be a problem, at the point the magnet is switched on, or the person is dropped into the magnet's helix, the change in magnetic field would induce significant electrical currents in the brain, similar to the experimental medical process transcranial magnetic stimulation, which can induce seizures.

Whether you are levitating a frog or a human being, though, this form of antigravity is still very limited as it only works within the confines of a fixed, ground-based magnet. The dream of antigravity researchers for many years is to cut the umbilical cord that connects us to our planet. They want to find some way to interfere with the Earth's gravitational pull, and as a result to be able to float over the ground without the need for jet packs, propellers, or other devices relying on Newton's third law of motion to push them away from the Earth.

On the Earth, were such antigravity possible, it would transform air travel, removing the need to exert energy in keeping a craft in the air and making an aircraft capable of executing sudden changes of direction that would rip a traditional aircraft apart. Similarly, if the antigravity process used relatively little energy, it would transform spaceflight, making it cheap to get objects into orbit or even to fly to the stars. It's not surprising that the possibility has proved attractive to serious flight engineers and crackpots alike. (To power a spacecraft, the antigravity device would have to go beyond simply blocking gravity and be able to manipulate the space-time around it.)

For a while there was considerable excitement about the possible uses of gyroscopes in antigravity. Anyone who has played with a gyroscope toy knows that they feel as if they really can counter the normal pull of gravity. They don't seem to act naturally. The best-known supporter of this concept was British electrical engineering professor Eric Laithwaite. In a lecture at the Royal Institution in London in 1974, Laithwaite claimed that a spinning gyroscope weighed less than it did when the flywheel was stopped.[2]

He demonstrated this using a heavy gyroscope at the end of a rod, which he could lift up with one hand if the gyroscope was spinning, but could not get off the ground when it was stationary. Originally Laithwaite claimed that gyroscopes operated outside Newton's laws of motion. Though he later retracted this suggestion, he continued to the end of his life to believe it should be possible to harness the power of gyroscopes to counter the effects of gravity.

Although there are still some devotees who believe Laithwaite had discovered true antigravity, all his demonstrations have since been recreated, with a clear description of the physics behind what is happening. A gyroscope can be used to maneuver a spacecraft in zero g, for example, because of the conservation of angular momentum. But a gyroscope is no more an antigravity device than is any other way of exerting a force to move something. It just goes about its business in a more subtle and confusing fashion.[3]

Over the years there have been a number of dedicated individuals who have put a large amount of effort into researching true antigravity. The devices they build are usually based on the assumption that general relativity is incorrect and that there is some linkage between gravity and electromagnetism. A typical example is the work of the early twentieth-century U.S. inventor Thomas Townsend Brown, who believed that high voltages could produce gravitational effects.[4]

Brown was born in Zanesfield, Ohio, in 1905 and started off on a fairly conventional scientific path at the California Institute of Technology, but seems to have been the kind of student who is either destined for great things or a career in fringe science.[5] He would get into arguments with the lecturers over what he could and couldn't experiment on. With enough family money to go it alone, he set up his own laboratory, where most of his "electro-gravitic" observations would be made, though he did spend time at two universities.

It ought to be stressed that Brown was by no means an out-and-out crank. He did legitimate work for the U.S. government between the First and Second World Wars that involved electromagnetism and radar. He was a competent electrical engineer. He based his ideas on science, not pseudoscience. But this does not mean that all his theories were necessarily correct.

It seems that Brown and those who followed in his footsteps, observed electromagnetic induction effects, rather like the comb lifting the pieces of paper, though much more powerful, driven by high-voltage electricity. These would certainly cause forces to be observed, but Brown assumed that these forces were caused by a modification of the pull of gravity. Brown's name has been given to a legitimate electromagnetic effect, the Biefield-Brown effect, which we will meet soon, but his speculation about the relation between electromagnetism and gravity had no good theoretical basis.

Whole schemes to replace general relativity have been built around the apparent results of this kind of experiment. Brown's electrogravitic ideas are not dissimilar to the work of Nikola Tesla, the Croatian/U.S. inventor who also denied the correctness of relativity and believed that powerful electrical charges were capable of acting outside the remit of traditional physics. Like Brown, Tesla knew his stuff on electromagnetism (among other things, Tesla invented the first practical AC electrical system), but this didn't prevent him from having extremely eccentric ideas in other areas.[6]

It usually isn't enough for those who want such antigravity technology to be real to stick to simple descriptions of experiments and theories. They inevitably suspect conspiracies. If, as they believe, such technology exists, then the government must be aware of it, and will make use of it. So, for instance, we see suggestions that the UFO-like B2 bomber uses antigravity technology to give it extra maneuverability, based on a difference between a positive electrical charge on its wings' leading edges and a negative charge applied to the jet exhaust.

There is some evidence that the B2 may make use of electrostatic technology in an attempt to increase stealth or reduce drag, but none at all that it employs antigravity technology.[7] Similarly, it is often suggested that flying disks powered by antigravity have been constructed in secret by the military ever since the Second World War, perhaps based on research carried out in Nazi Germany. The story goes that these craft are responsible for the prevalence of stories of disk-shaped flying saucers.

This particular conspiracy theory is highly unlikely to have any merits. Apart from anything else, the term "flying saucer" came from a newspaper report of a sighting of some unusual craft by U.S. pilot Kenneth Arnold in 1947. Arnold did not say at the time that the craft he saw were saucer-shaped. Instead he said that they moved erratically, "like a saucer if you skip it across the water."[8] The term was coined by newspaper headline writers and readers mistakenly assumed that the "saucer" part referred to the shape of the craft. Sightings of saucer-shaped UFOs started shortly after the term started to be used in the press.

It is true that flying disk aircraft have been developed on and off ever since the 1930s. These largely depended on a conventional propeller engine directed downward in the center of a ring-shaped craft, but they have never proved particularly controllable or stable in flight and have remained novelties rather than practical flying machines.

These flying disks were conventionally powered, but the enthusiasm for the potential for antigravity continued unabated. In the 1950s a popular engineering magazine published an article full of the possibilities that the writer believed were soon to emerge from the antigravity research that surely had to be going on in secret. "Future aircraft will attain the speed of light," the headline blared. ". . . scientists, designers, and engineers are perfecting a way to control gravity—a force infinitely more powerful than the mighty atom. The result of their labors will be anti-gravity engines working without fuel—weightless airliners and space ships able to travel at 170,000 miles per second."[9]

It's hard to know where to start on what is wrong with this statement. Gravity is not "infinitely more powerful than the mighty atom"—it is by far the weakest of the four forces of nature, trivially weak compared with the mighty atom. Even if they could be built, no electrostatic antigravity device like Townsend

Brown's could work without fuel—it would require a large amount of electrical energy to operate. But strangest of all is the bizarre assertion that once severed from gravity, objects would travel at practically the speed of light.

This seems to demonstrate a lack of awareness of basic Galilean relativity. If something is weightless it would float up in the atmosphere like a hot air balloon—but there is no reason why it should shoot off at extremely high speeds. Nor is there any particular reason to assume that because something is weightless it also lacks mass—a totally separate concept. If a ship had a sizeable mass, it would be difficult to provide enough energy to get it anywhere near the speed of light. This is just confused thinking.

To add more confusion to the mix, there have been legitimate scientists who have devised flying machines with more than a passing resemblance to Thomas Townsend Brown's concepts, but who are aware that the only force acting against gravity is electromagnetism, not some product of the creative but fictional electrogravitic force of Brown's imagination. For example, Professor Subrata Roy of the University of Florida released details in 2008 of a flying saucer that would become airborne without moving parts.[10]

The apparent antigravity of Roy's saucer plans are just a variant on the electromagnetic levitation we've already seen. The saucer would be covered with electrodes that create sufficient electric charge to convert the air under the ship into plasma—a gas of charged particles. At this point standard electromagnetic repulsion is used to enable the ship to repel the plasma, pushing it away from the ground and making it hover. This is the Biefield-Brown effect, named after Townsend Brown.

This approach is also responsible for the equipment often claimed to demonstrate antigravity in action, known as "lifters." These are flimsy foil—constructed devices, often triangular in shape, which are given an electrical charge to produce ionized air (a partial plasma) beneath them, which then produces lift as it is repelled by the charged device.

Although conspiracy theorists delight in the idea that somewhere, top secret defense establishments have produced working antigravity technology, it seems highly unlikely that it is true. There would be too many commercial uses, making too many billionaires, for the secret to be hidden away completely. And there is no parallel with known breakthrough technologies. The jet engine, for example, was impossible for the military to keep secret to further national advantage. There is no reason why they should be more successful with antigravity and every reason that such an amazing technology would soon leak out.

What's more, even if the specific devices are classified, the science that lies behind the secret technology rapidly becomes widely known. Take nuclear weapons, for example. The knowledge required to make nuclear weapons has always been highly classified, as is the detailed knowledge of the technology to make their construction possible. Yet the science behind them was already widely known before the bomb was first constructed. It is very hard to gag science in the way that conspiracy theorists would have us believe that the military/industrial complex is capable of doing.

Perhaps the best example to suggest that it is highly unlikely that antigravity technology has been kept secret for 50 years, is the way that a known super-secret technology, stealth technology, has come into the public domain. The U.S. military has spent billions of dollars on the stealth technology that went into aircraft like the B2 and the F117-A. These developments were what is sometimes referred to as "black developments," off-the-balance-sheet expenditure that wasn't reported in open government accounts.

Yet the fact remains we all now know about stealth technology, which has no great commercial value outside of the military. Antigravity—which we don't see deployed in battlefields around the world—is on a whole different scale of commercial importance. It would make vast amounts of money if it were made commercial. Such a technology simply could not be kept locked away indefinitely if there were any real benefit behind the hype and military intelligence smokescreens.

Those who believe antigravity technology has been around for many years often point back to secret research undertaken by the Germans toward the end of the Second World War, research producing technology that is thought to have been captured by the United States in raids immediately after Germany fell. We know that this is how the United States obtained a lead in rocket expertise when Wernher von Braun, other scientists, and their equipment were brought across the Atlantic. But we have to be wary of believing every claim of amazing technology of the time.

Books suggesting that the Nazis had antigravity technology also often claim that they had "anti-aircraft rays."[11] Leaving aside the fact that they didn't seem to have succeeded in destroying too many enemy aircraft with rays, something they surely would have done had the weapons existed, this provides a powerful insight into the most likely status of any antigravity project. The insight comes from the early history of radar.

Originally called "range and direction finding" before the name was changed to "radio detection and ranging," which was then shortened to radar, the first known work on the technology was performed in the British Radio Research Station at Slough, England.[12] The experts there had been asked by their government to investigate claims from Nikola Tesla he had constructed a death ray using powerful electromagnetic radiation.

The scientists at Slough could see no way to use radio to kill people, but they did discover the effectiveness of electromagnetic beams for detecting objects like aircraft. It would be easy enough for reports to leak out that scientists in England had developed prototype electromagnetic beam weapons, because it *was* a technology they investigated. The fact that they could make nothing of the idea didn't mean they hadn't tried. This is likely to be the case with the German anti-aircraft rays . . . and their antigravity technology. It was an earnest but doomed attempt to duplicate spurious effects that then gets reported as incredible advanced technology.

We do occasionally still see scientists, rather in the manner of Tesla and Townsend Brown, observing something which they ascribe to antigravity, but which then proves impossible to duplicate elsewhere. Most recently, the best publicized example is that of Russian engineer Yevgeny Podkletnov, who claims to have observed antigravity effects from a rotating superconductor.[13]

Podkletnov's is a much more hi-tech possibility than Laithwaite's gyroscopes. His device combines some of the most bizarre-known phenomena in physics with engineering expertise that has stretched the capabilities of those who have tried to reproduce his work.

A superconductor is a material that has no electrical resistance, allowing current to flow through it without restraint. In principle, once started, a current could flow around a ring of superconductor forever. Superconductivity is a quantum effect that requires very low temperatures, although so-called high-temperature superconductors can operate as "hot" as –180 degrees Celsius (–292 degrees Fahrenheit).

Superconductivity can itself produce a form of apparent antigravity. One of the defining features of a superconductor is that it excludes any magnetic field from within the superconductor as it forms, so that a magnet placed on top of a superconductor will begin to "levitate" and float above it. However Podkletnov claimed to have seen something more.

In the 1990s, working with a high-speed rotating superconductor, he claimed that objects above the superconductor (both metals and insulators) experienced a reduction in weight. This was nowhere near a total countering of the Earth's gravitational field, but a reduction of around 2 percent. Despite claims from Podkletnov that his experiments had been reproduced elsewhere, there seems to be no such reproduction and he was soon excluded from his laboratory and dropped out of the antigravity business.

It has been suggested that a weak effect might exist due to the gravitomagnetic component of general relativity (see page 153), in effect a form of frame dragging from the rotating disk, but this has not been confirmed. There were claims in 2002 that a number of bodies from NASA to Boeing and BAE Systems were looking into this effect to see if it could have practical uses, but none of these reports were followed up by news of discoveries.[14] It seems the Podkletnov effect, if it existed at all, was a dead end. Most likely it was an electromagnetic effect, like the levitating frog, being mistaken for gravitation.

The most extreme examples of antigravity pseudoscience link the ability to reduce gravitational pull with zero-point energy. This is a real physical phenomenon that has built up an amazing panoply of pseudoscience around it because it sounds incredible, but it is generally considered to be much less useful than the antigravity enthusiasts seem to think.

Zero-point energy is another term for vacuum energy. It is based on the idea that even totally empty space isn't empty. Quantum theory tells us that there are certain pairs of properties that are linked—the more we know about one, the less we know about the other. This is Heisenberg's uncertainty principle, best known for linking momentum and position. But it also applies to energy and time.

What this tells us is that in a very precise bit of time, the amount of energy in a section of space will inevitably be very poorly defined. It has the potential to be very high, high enough for the energy to convert into a matter/antimatter pair of particles. So empty space is predicted to be seething with these virtual particles popping into existence, colliding and annihilating, never existing long enough to be detected.

The net energy over a large area of space is zero, but isolate a small part of space in the right way and this effect means that you can lay your hands on some negative energy. This is the well-documented Casimir effect, first proposed by the Dutch physicist Hendrik Casimir. Put two flat plates very, very close together and it will not be possible for virtual particles to form between them, because the distance is smaller than the wavelength of the probability wave that defines the position of the particles in quantum theory. There will be virtual particles forming outside the plates but not inside. Because there is pressure from the particles outside, but none inside, the net result is a negative energy between the plates which pulls them together.

Antigravity enthusiasts suggest that antigravity devices like Podkletnov's tap into the apparently limitless resources of zero-point energy. But anyone intending to use zero-point energy for power seems to have forgotten the laws of thermodynamics. To get energy out of something you need a "sink"—somewhere else that has less energy. Take a simple example. If I put a ball on top of a mountain and let go, it will roll down because there is a sink, a point of lower energy. If everywhere around the starting point had the same potential energy or more, the ball would go nowhere.

With a heat engine, the sink is usually the temperature of cold water, though to get maximum efficiency out of a heat engine you really want to get that sink as close to absolute zero as possible. The maximum energy you can get out depends on the difference between your operating energy and that of the sink. Again, if there is nowhere with lower energy than your starting point, you can't get energy out.

So to harness zero-point energy you would need a sink that had lower energy than zero point—which by definition is the minimum achievable energy. (Casimir effect plates are a special case because they manage to circumvent zero-point energy, but only on a very small scale.) Schemes to extract free energy from zero-point energy are attempting the impossible. The energy is there, but you can't make practical use of it.

Back in the real world of accepted physics, there is one possibility for a material that does in some ways act contrary to normal matter. This could be a way to produce a substance that responds negatively to a gravitational field, treating it as repulsion instead of attraction. After all, there are two different kinds of matter—normal matter and antimatter.

Antimatter is usually described as being the same as ordinary matter but with an opposite charge. So the antimatter version of an electron, a positron (or antielectron) is identical to an ordinary electron, but has a positive electrical charge instead of a negative charge. But this isn't the whole story. There are, for example, antineutrons. These particles have no electrical charge, but are still antimatter equivalents of ordinary neutrons, still different despite having the same charge.

Although antineutrons were known before the subcomponents of particles like neutrons had been discovered, we now know that an ordinary neutron is made up of one up quark and two down quarks (with charges $+\frac{2}{3}$ and $-\frac{1}{3}$ respectively) while an antineutron is made up of one up antiquark and two down antiquarks (charges $-\frac{2}{3}$ and $+\frac{1}{3}$).

The quarks also have other properties reversed, and it is just possible that one of the properties that is switched is gravitational mass. It might seem that this should have been easily tested, but we have only ever had very, very small amounts of antimatter, which typically exist for a tiny fraction of a second before the antimatter interacts with ordinary matter and transforms to energy.

Because the gravitational force is so weak, the amount of force acting on, say, a single antineutron is ludicrously tiny, and the displacement under the influence of gravity that can take place in their lifetime to date is not measurable. In principle antimatter could have a negative gravitational mass, though most physicists doubt this is likely. But they do not rule out antimatter having a different response to gravity to conventional matter, which would have the same effect as a negative mass.[15]

Specifically, according to Italian astrophysicist Massimo Villata, although showing that matter should attract matter and antimatter should attract antimatter gravitationally, general relativity predicts that matter and antimatter should repel each other.[16] This does not involve antimatter having a negative mass, although the outcome is as if it had. The mass going into the equations is still positive, but other factors (equivalent to the particle's position and momentum) are reversed, resulting in antimatter behaving as if it did have negative gravitational mass.

In this scenario, charged matter and antimatter particles like an electron and a positron would still be attracted electromagnetically to each other and annihilate, because the electromagnetic force is much stronger than the antigravity repelling them, but it does offer a small possibility that if stable antimatter atoms could be produced they would provide a true form of antigravity.

Antihydrogen—hydrogen atoms with a positron orbiting an antiproton—have been produced at CERN, but there is a serious problem if uncharged antimatter is to be used in any kind of vehicle. While it is possible to contain charged antimatter particles in a magnetic bottle, where the electromagnetic field holds it in place, there is no way to keep uncharged atoms from bumping into nearby matter and being destroyed.

If such antimatter does generate antigravity, then it would provide a repulsive force against the walls of the container, but not enough to overcome collisions as the atoms shoot around inside. There is also the minor problem that at the moment the world's annual production of antimatter is measured in millionths of a gram—not an amount that would be enabling an antigravity craft to fly any time soon.

One interesting side effect of this development, were it true, is that the principle of equivalence which first inspired general relativity would have to be modified.[17] It would no longer be the case that you could not tell the difference between acceleration and gravity. Under acceleration, both matter and antimatter should move in the same direction, but under gravity they would move in opposite directions.

This isn't particularly a problem for general relativity—it could be allowed for—but equivalence would become a less universal principle. It would still be fine for pure matter or pure antimatter worlds—and still provide the inspiration behind general relativity—but where a mix of the two types of matter existed it would have to be discarded.

General relativity tells us that one of the components of gravity is pressure. When gravity was purely dependent on mass, in Newton's formulation, the concept of negative gravity was pretty meaningless, as, unless antimatter demonstrates otherwise, we have never discovered anything with a negative mass. But pressure is a different animal. It is entirely possible to have negative pressure.

This might seem unlikely, but just think of what pressure is. Imagine I've got a wall and I fire a continuous bombardment of tennis balls at its left-hand side. This puts the wall under pressure, though we are more used to pressure being due to the continuous bombardment of the billions of molecules in a gas.

The effect of the bombardment, the pressure, is to push the wall to the right, thanks to Newton's third law of motion.

Now let's imagine we fix a large elastic band to the left-hand side of the wall and stretch that band away from the wall. The effect will be the opposite of the pressure—instead of pushing the wall to the right it will pull it to the left. This is negative pressure. It's tension in the band that produces the effect. Tension is the equivalent of negative pressure. And if we can introduce enough negative pressure into a situation that is producing gravity, with low enough mass, the outcome should be negative gravitational force.

A form of negative pressure entered general relativity at an early stage, was thrown out, and then returned as reality proved more interesting than theory. Look at Einstein's field equations again:

$$G_{\mu\nu} + \Lambda g_{\mu\nu} = (8\pi G/c^4)\ T_{\mu\nu}$$

The constant Λ was originally put in as a fudge factor. As far back as Newton, there had been an awareness that the universe should not be stable. All the matter in the universe should be attracting all the other matter, giving the whole thing a tendency to collapse in on itself. It might take a long time, but the universe should coalesce. Newton decided that the only way to avoid this was if the universe was infinite, so that every bit of matter in the universe was attracted from all sides and so stayed in place. There would be no center for everything to head toward.

This was a very delicately balanced picture. After all, very little in the universe seemed to be absolutely static, yet the slightest movement out of position would render the system unstable and parts of it would collapse. An infinite universe wouldn't all head for the same place, but would have multiple points where everything was crushed together. Newton's only solution was to call on God to act as a universal correcting system, pushing bodies back into place if they were in danger of starting a collapse.

For Einstein, God wasn't an option, so he introduced an opposing force to the gravitational pull, a negative pressure in the universe that exactly balanced the natural tendency to collapse. This was that value Λ, often called the cosmological constant. Einstein later referred to this as his greatest mistake, because it was put in purely to fudge things. There was no reason for it to be there, no known natural force that produced this kind of negative pressure, it merely stabilized the universe.

At the time, Einstein had no conception of an expanding universe. He expected the size of the universe to be constant, so Λ was there to keep things balanced. Observations proved him wrong. Not only is the universe expanding, but that expansion is accelerating. The cosmological constant is back in a big way to represent dark energy, the driving force behind the acceleration of the universe. And another way of looking at dark energy is that it is negative pressure.

Remember the picture of the elastic band, pulling on the wall. Dark energy is the elastic band that pulls on the universe to make it ever larger. It seems that dark energy—something that is, in effect, antigravity—makes up around 70 percent of the universe. At first sight this seems to imply that there is plenty of this antigravity force to go around, so we should be able to employ it to create an antigravity device. And as it appears to extend throughout the universe, why not harness it as we go with an antigravity ship?

There are two significant problems. One is just how to harness it. We know it's there, because we can see the effect it causes. But there is no antigravity socket in the sky to plug into. There is no obvious way to make use of it. But even if there were, at the local level, dark energy is so insignificant that it has little benefit. Just because it makes up 70 percent of the universe, doesn't make its local effect large—the universe is a big place.

We have a large reminder of this in the sky, in the constellation of Andromeda. As we've already seen, the Andromeda galaxy is heading toward us because the gravitational attraction between that galaxy and the Milky Way is stronger than the expansion of the universe. The Andromeda galaxy is 2.5 million light-years away, and yet over that range, weak old gravity is still winning. Even if we could harvest dark energy as an antigravity power source, there wouldn't be enough in the vicinity to have a useful effect.

For the moment, then, antigravity seems to be best consigned to science fiction and to the conspiracy theorists. If it shows us anything it is just how important gravity is to us, when we are prepared to exert so much effort in a futile attempt to counter it.

CHAPTER TWELVE

CENTER OF ATTRACTION

> *But the most impressive fact is that gravity is simple. It is simple to state the principles completely and not have left any vagueness for anybody to change the ideas of the law. It is simple and therefore it is beautiful. It is simple in its pattern. I do not mean it is simple in its action—the motions of the various planets and the perturbations of one on the other can be quite complicated to work out, and to follow how all those stars in a globular cluster move is quite beyond our ability. It is complicated in its actions, but the basic pattern, or the system beneath the whole thing is simple.*
>
> —*The Character of Physical Law* (1967)
>
> Richard Phillips Feynman

Gravity may be the weakest of the fundamental forces, but it's gravity that provides many of the everyday challenges we face. We are fighting gravity each time we take a step, and gravity helps encourage the gradual collapse of various parts of our anatomy. As we come to understand gravity better, we may be able to exert more control over it, just as we do over some electromagnetic forces, but we can never take it for granted.

When Richard Feynman, without doubt one of the greatest American scientists of all time, wrote that the impressive thing about gravity was that it was simple, he was not showing off. Even to Feynman, the tenfold equations hidden behind Einstein's deceptively simple field equations were challenging. But Feynman had a refreshingly honest way to look at complicated science.

He happened to be talking about quantum theory when he said the following in a lecture in 1983, but he could have said it just as easily about general relativity and all the attempts to create a quantum theory of gravity:[1]

> What I am going to tell you about is what we teach to our physics students in the third or fourth year of graduate school—and you think I'm going to explain it to you so you can understand it? No, you're not going to be able to understand it. Why, then, am I going to bother you with all this? Why are you going to sit here all this time, when you won't be able to understand what I am going to say? It is my task to convince you not to turn away because you don't understand it. You see, my physics students don't understand it either. That is because *I* don't understand it. Nobody does.

Feynman's paean to not understanding was not the frustrated cry of someone giving up. He was not saying, "We don't understand it, so there's no point trying." Instead he was trying to counter some of the reasons that we turn away from "difficult" science. He was explaining first of all that physics is never going to be able to answer the ultimate question: "Why?" We can describe how nature behaves with more and more accuracy. We can observe apparent laws and constants of nature. But we can never answer questions like, "Why does gravity attract?"

Some try to deal with this kind of uncertainty by speculating that there might be many different universes in which every possible value or characteristic of nature is different. So there might be a universe where G is much bigger, or smaller, or varies by day of the week. There might be a universe where gravity comes with opposing poles, like a magnet, or where gravity doesn't exist at all. With such a multiverse available to us, they have a simple answer to why our universe is the way it is. It's because we just happen to be in that universe.

Actually, their response may well be a little stronger, invoking the anthropic principle. If we assume such a set of different universes exists, we could only ever inhabit a universe where atoms can form, planets can form, stars can form, and so on. Without that happening we wouldn't be here. And that requirement limits the possibilities for variation in the different forces and constants. As we have already seen, if gravity were much stronger, the universe would have collapsed back in on itself. Similarly if the strong nuclear force were a little weaker, the nuclei of atoms would not hold together.

So given the multiverse, the simple answer to "Why are things like they are?" is that they have to be or we wouldn't be here.

Many people (including me) find this answer facile. Even if you accept it, all we've done is push back the "Why" a stage to ask instead, "Why does that multiverse exist?" At this level we move away from science and get to metaphysics or theology.

On one level, then, Feynman was saying that there has to be a point at which science can't meaningfully answer the question, "Why?" But he also had a second point in mind. This was the possibility that on hearing about something, you just can't believe it. In effect you don't like what you hear, so you reject it. As he says:[2]

> It's a problem that physicists have learned to deal with: They've learned to realize that whether they like a theory or they don't like a theory is *not* the essential question. Rather it's whether or not the theory gives predictions that agree with experiment. It is not a question of whether a theory is philosophically delightful, or easy to understand, or perfectly reasonable from the point of view of common sense.

This is something that is equally valuable advice to those who devise abstruse mathematical formulations to attempt to tame quantum gravity and for those of us struggling with understanding something like general relativity. Everyone working on string theory, or loop quantum gravity, or their rivals should be issued with a poster with Feynman's words printed in large letters. For them, the key phrase is that "It is not a question of whether a theory is philosophically delightful," a trap that theorists often fall into.

For the rest of us, it is the other two provisos that are particularly important. The concepts behind the formulae of general relativity are not easy to understand. The whole idea that gravity is a result of warped space-time is not reasonable from the point of view of common sense. But the fact is, for the moment at least, general relativity is the best theory we've got in terms of giving predictions about gravity that agree with experiment.

Along with Richard Feynman, we all need to let the difficulties wash over us, to put the attractive but ultimately false nagging of common sense to one side. Then we can see just how wonderful gravity is in everything it does, and how remarkable is general relativity's ability to explain the inexplicable.

When I first came across general relativity, the most wonderful moment was the understanding of the significance that not only space, but time, too, was warped. I had struggled with the rubber sheet model, wondering why things started to move when attracted by gravity, just because space was distorted. But the idea that time is warped, too, by a massive object, making a passage through time also a passage through space, was a revelation.

Taken further still, this warping of space and time leads to all kinds of wonderful possibilities like time travel and wormholes in space. And we can have great fun playing around with the potential for antigravity devices, and all the glorious conspiracy theories that surround those. But in the end, for me, what is most delightful is that simplicity that Feynman identified. It crops up first in Newton's discovery of the scale of the gravitational force and then in Einstein's replacing that mystical force at a distance with the satisfying solidity of warped space-time.

Pick up a book. Drop it. Watch gravity in action. And every day you can see how the weakest force in the universe shapes our lives.

NOTES

CHAPTER 1

1. The experiment with quail eggs sponsored by KFC is described in Michael Brooks, "Gravity Mysteries: does life need gravity," *New Scientist,* June 10, 2009.

CHAPTER 2

1. The possible attribution of the spherical Earth idea to Pythagoras is mentioned in James Evans, *The History and Practice of Ancient Astronomy* (Oxford: Oxford University Press, 1998), p. 47.
2. The tendency to ascribe ideas to Pythagoras he wasn't responsible for (including his theorem) is described in Kitty Ferguson, *Pythagoras: His Lives and the Legacy of a Rational Universe* (London: Icon Books, 2011), p. 76ff.

3. Plato's assertion that the Earth is like a leather ball is from Plato, *Phaedo* (trans. Benjamin Jowett, text from en.wikisource.org/wiki/Phaedo).
4. Empedocles' theory of the four elements is described in Brian Clegg, *Light-Years* (London: Macmillan Science, 2007), p. 16.
5. The idea that gravity doesn't exist and the effect we describe as the gravitation is in fact the expansion of matter is the basis of Mark McCutcheon, *The Final Theory: Rethinking Our Scientific Legacy* (Boca Raton: Universal Publishers, 2010), all pp.
6. The concept of objects behaving a particular way because of their nature, rather than mechanical components in Aristotle's physics is described in Elspeth Whitney, *Medieval Science and Technology* (London: Greenwood Press, 2004), p. 48.
7. Aristotle's argument that air continued the push to keep a thrown object moving against its natural tendency to fall is described in Elspeth Whitney, *Medieval Science and Technology* (London: Greenwood Press, 2004), p. 55.

CHAPTER 3

1. The medieval movement from the ancient Greek qualitative approach to science to a more quantitative approach is described in Elspeth Whitney, *Medieval Science and Technology* (London: Greenwood Press, 2004), p. 59.
2. Information on Roger Bacon is from Brian Clegg, *The First Scientist* (London: Constable & Robinson, 2003), all pp.
3. Roger Bacon's assertion that experiment is essential to under-

stand natural phenomena is from Roger Bacon (trans. Robert Belle Burke), *The Opus Majus* (Philadelphia: University of Philadelphia Press, 1928; reprinted Kessinger Publishing, 1942), p. 584.

4. Roger Bacon's emphasis on the importance of mathematics to the sciences is from Roger Bacon (trans. Robert Belle Burke), *The Opus Majus* (Philadelphia: University of Philadelphia Press, 1928; reprinted Kessinger Publishing, 1942), p. 116.

5. The medieval idea that the planets were powered by angels is from Elspeth Whitney, *Medieval Science and Technology* (London: Greenwood Press, 2004), p. 70.

6. John Buridan and Nicole Oresme's idea that it made more sense for the Earth to rotate in 24 hours than the universe is described in Elspeth Whitney, *Medieval Science and Technology* (London: Greenwood Press, 2004), p. 72.

7. Adelard of Bath's description of the workings of gravity is from Adelard of Bath (trans. Charles Burnett), *Adelard of Bath, Conversations with his Nephew* (Cambridge: Cambridge University Press, 1998), p. 181.

8. Peter Peregrinus and his *De magnete* are described in Brian Clegg, *The First Scientist* (London: Constable & Robinson, 2003), p. 32.

CHAPTER 4

1. Information on the context of Archimedes' book *The Sand Reckoner* is from Brian Clegg, *A Brief History of Infinity* (London: Constable & Robinson, 2003), pp. 37–44.

2. The quote from Archimedes about Aristarchus and the Sun-centered universe is from Archimedes (ed. T. L. Heath), *The Works of Archimedes—The Sand Reckoner* (New York: Dover Publications, 2002), pp. 221–22.

3. Information on Copernicus is from Brian Clegg, *Light-Years* (London: Macmillan Science, 2008), pp. 46–50.

4. Biographical information on Galileo is from Brian Clegg, *Light-Years* (London: Macmillan Science, 2008), pp. 57–62.

5. Galileo's distaste with his book's new title is quoted by Antonio Favaro in the introduction to Galileo Galilei (trans. Henry Crew and Alfonso de Salvio), *Dialogues Concerning Two New Sciences* (New York: Dover, 1954), p. xii.

6. Galileo's surprise that his book had been published when all he was doing was showing some ideas to a few friends is in Galileo Galilei (trans. Henry Crew and Alfonso de Salvio), *Dialogues Concerning Two New Sciences* (New York: Dover, 1954), pp. xvii–xviii.

7. Galileo's refutation of Aristotle's idea that a heavy object falls faster than a light object is in Galileo Galilei (trans. Henry Crew and Alfonso de Salvio), *Dialogues Concerning Two New Sciences* (New York: Dover, 1954), pp. 62–63.

8. The video of Scott dropping a hammer and feather on the Moon can be seen at http://video.google.com/videoplay?docid=6926891572259784994.

9. Galileo's description of the action of pendulums with different weights is in Galileo Galilei (trans. Henry Crew and Alfonso de Salvio), *Dialogues Concerning Two New Sciences* (New York: Dover, 1954), pp. 84–86.

10. The inaccuracy in Galileo's idea that simple pendulums swing with the same period whatever their amplitude is described in Gregory L. Baker, *Seven Tales of the Pendulum* (Oxford: Oxford University Press, 2011), pp. 111–12.

11. The use of pendulums to measure the acceleration due to gravity is described in Gregory L. Baker, *Seven Tales of the Pendulum* (Oxford: Oxford University Press, 2011), pp. 25–27.

12. Galileo's introduction to his "new science" is from Galileo Galilei (trans. Henry Crew and Alfonso de Salvio), *Dialogues Concerning Two New Sciences* (New York: Dover, 1954), p. 153.

13. Galileo's statement of the scientific method is from Galileo Galilei (trans. Henry Crew and Alfonso de Salvio), *Dialogues Concerning Two New Sciences* (New York: Dover, 1954), p. 160.

14. Galileo's stated fallacy about acceleration due to gravity being proportional to distance traveled is from Galileo Galilei (trans. Henry Crew and Alfonso de Salvio), *Dialogues Concerning Two New Sciences* (New York: Dover, 1954), p. 167.

15. Biographical information on Newton is from Brian Clegg, *Light-Years* (London: Macmillan Science, 2008), pp. 73–89.

16. Steve Jobs's origination of the company name Apple is described in Steven Levy, *Hackers* (London: Penguin, 1994), p. 253.

17. Isaac Newton's apple incident is recounted from the manuscript version of William Stukeley, *Memoirs of Sir Isaac Newton's Life* (London: Royal Society, 1752), at royalsociety.org/turning-the-pages/ p. 15ff.

18. The damage caused to Newton's apple tree by tourists is

described in "Gravity of damage facing Newton's tree prompts action," *The Times* (London), May 12, 2011.

19. Unless otherwise specified, information on Newton's *Principia* is from Isaac Newton (trans. J. Bernard Cohen and Anne Whitman), *The Principia—Mathematical Principles of Natural Philosophy* (Berkeley and Los Angeles: University of California Press, 1999).

20. Newton's discovery of Huygens's rule for the force on a body traveling a circle is described in John Herivel, *The Background to Newton's "Principia": A Study of Newton's Dynamical Researches in the Years 1664–84* (Oxford: Clarendon Press, 1965), p. 130.

21. Newton's discussion with Robert Hooke on Hooke's "hypothesis" on orbital motion is covered in a range of letters in volume 2 (1676–1687) of Isaac Newton (ed. H. W. Turnbull), *The Correspondence of Isaac Newton* (Cambridge: Cambridge University Press, 2008).

22. The discovery that weight varies in different parts of the world by Halley and others is described in Isaac Newton (trans. J. Bernard Cohen and Anne Whitman), *The Principia—Mathematical Principles of Natural Philosophy* (Berkeley and Los Angeles: University of California Press, 1999), p. 87.

23. Newton's instruction to the reader on how to skip through parts of the book are from Isaac Newton (trans. J. Bernard Cohen and Anne Whitman), *The Principia—Mathematical Principles of Natural Philosophy* (Berkeley and Los Angeles: University of California Press, 1999), p. 793.

24. The establishment of the universal gravitational constant by

Charles Vernon Boys is described in Gregory L. Baker, *Seven Tales of the Pendulum* (Oxford: Oxford University Press, 2011), p. 98.

25. Experiments using torsion pendulums to measure the gravitational constant are described in Gregory L. Baker, *Seven Tales of the Pendulum* (Oxford: Oxford University Press, 2011), pp. 99–103.

CHAPTER 5

1. The explanation of the implication of Newton's term "fingo" (frame) is from the notes to Isaac Newton (trans. J. Bernard Cohen and Anne Whitman), *The Principia—Mathematical Principles of Natural Philosophy* (Berkeley and Los Angeles: University of California Press, 1999), p. 275.

2. Huygens's dismissal of Newton's basis of theories on the principle of attraction is from a letter from Huygens to Leibniz in Christiaan Huygens, *Oeuvres complètes, publiées par la Société hollandaise des sciences* (The Hague: Martinus Nijhoff, 1888–1950), v. 9, p. 190.

3. The comments by Leibniz on Newton's use of a gravitational force of attraction are from Gottfried Wilhelm Leibniz, *Die philosophischen Schriften* (Berlin: Weidmann, 1875–1890), v. 5, p. 58.

4. Newton's letter to Richard Bentley is described in Isaac Newton (trans. J. Bernard Cohen and Anne Whitman), *The Principia—Mathematical Principles of Natural Philosophy* (Berkeley and Los Angeles: University of California Press, 1999), p. 62.

5. Information on the corpuscle-based mechanical explanation of

gravity is from William Bower Taylor, *Kinetic Theories of Gravitation* (Annual Report of the Board of Regents, Smithsonian Institution, v. 31 (1876), pp. 205–82), reproduced at en.wikisource.org/wiki/Kinetic_Theories_of_Gravitation.

6. Lord Kelvin's contribution to the Le Sage theory is from William Thomson, *Mathematical and Physical Papers: 108—On the Ultramundane Corpuscles of Le Sage* (University of Michigan Historical Math Collection, from the Proceedings of the Royal Society of Edinburgh v. VII (1872), pp. 577–89), reproduced at quod.lib.umich.edu/cgi/t/text/pageviewer-idx?idno=aat1571.0005.001&c=umhistmath&seq=81.

7. Information on specific examples of injury is from "celebratory gunfire" and of tests undertaken to see if this is a real hazard is provided on Wikipedia at en.wikipedia.org/wiki/Celebratory_gunfire.

8. Galileo's discussion of the Moon's role in the tides is from Galileo Galilei (trans. Stillman Drake), *Dialogue Concerning The Two Chief World Systems* (New Berkeley: University of California Press, 1953), pp. 419–20.

9. Roger Bacon's description of how the Moon causes the tides is from Roger Bacon (trans. Robert Belle Burke), *The Opus Majus* (Philadelphia: University of Philadelphia Press, 1928; reprinted Kessinger Publishing, 1942), pp. 161–62.

10. Newton makes several references to the tides, but his main description of the influence of gravitation from the Sun and Moon is from Isaac Newton (trans. J. Bernard Cohen and Anne Whitman), *The Principia—Mathematical Principles of Natural Philos-*

ophy (Berkeley and Los Angeles: University of California Press, 1999), p. 835ff.

11. The detailed explanation of the tides is based on George Gamow, *Gravity* (New York: Dover, 2002), pp. 81–84.

12. The explanation of the Moon always keeping one face toward the Earth due to tidal forces is from Bernard Schutz, *Gravity from the Ground Up* (Cambridge: Cambridge University Press, 2003), pp. 44–45.

13. The description of tidal forces and their effect on the moons of Jupiter is from Bernard Schutz, *Gravity from the Ground Up* (Cambridge: Cambridge University Press, 2003), p. 46.

14. Details of Herschel's discovery of Uranus is from Michael Hoskin, *Discoverers of the Universe—William and Caroline Herschel* (Princeton: Princeton University Press, 2011), pp. 49–50.

15. Information on quasars is from Bernard Schutz, *Gravity from the Ground Up* (Cambridge: Cambridge University Press, 2003), pp. 176–77.

16. Details of the Goce gravitational satellite are from the European Space Agency (http://www.esa.int/esaLP/LPgoce.html) and the BBC (http://www.bbc.co.uk/news/science-environment-12911806).

CHAPTER 6

1. Biographical information on Einstein is taken from Michael White and John Gribbin, *Einstein—a Life in Science* (London: Pocket Books, 1997), all pp.

2. Einstein's original paper on special relativity, "Zur Elektrodynamik bewegter Körper," was published in *Annalen der Physik*. 17:891, 1905.

3. Einstein's memory of his inspiration for general relativity while sitting in the Bern patent office is from his 1922 Kyoto Lecture and referenced in W. F. Bynum and Roy Porter (eds.), *Oxford Dictionary of Scientific Quotations* (Oxford: Oxford University Press, 2005), p. 198.

4. "The happiest thought of my life" comment was originally made in an unused draft of an article for *Nature*, on general relativity, quoted in Abraham Pais, *Subtle Is the Lord* (Oxford: Oxford University Press, 1983), p. 178.

5. The time traveler's description of time as a fourth dimension is from H. G. Wells, *The Time Machine* (London: Pan Books, 1965), p. 8.

6. The invariance of distances in space-time are described in Brian Cox and Jeff Forshaw, *Why Does $E=mc^2$?* (Cambridge: Da Capo Books, 2009), p. 64.

7. The analogy comparing attraction in general relativity with walking to the North Pole is from Brian Cox and Jeff Forshaw, *Why Does $E=mc^2$?* (Cambridge: Da Capo Books, 2009), p. 227.

8. Arthur Eddington says "let us represent spherical space by a rubber balloon" in Arthur Eddington, *The Expanding Universe* (Cambridge: Cambridge University Press, 1933), p. 66.

9. I would rather not identify the professors who didn't come up with an explanation at all, or complained there was nothing to understand because a force was just a force, as it might seem

derogatory. They are, however, listed in the acknowledgments along with other more constructive helpers.

10. Newton's query on the bending of light by gravitation in the third volume of *Opticks* is referenced in Brian Clegg, *Light-Years* (London: Macmillan Science, 2007), p. 90.

CHAPTER 7

1. Erwin Freundlich's and Eddington's expeditions to attempt to provide data to test the general theory of relativity are described in Brian Clegg, *Light-Years* (London: Macmillan Science, 2007), pp. 184–85.

2. Details of total solar eclipses from 1914 to 1919 are taken from Fred Espenak and Jean Meeus, *Five Millennium Catalog of Solar Eclipses:* –1999 to +3000 (2000 BCE to 3000 CE)—Revised, NASA/TP-2009-214174 (http://eclipse.gsfc.nasa.gov/5MCSE/TP2009-214174.pdf), p. A-158.

3. The details of Eddington's Principe observation are in Arthur Eddington, Frank Dyson, and Charles Davidson, "A Determination of the Deflection of Light by the Sun's Gravitational Field, from Observations made at the Total Eclipse of May 29, 1919," Philosophical Transactions of the Royal Society of London series A, 220, pp. 291–333, 1920.

4. The dubious nature of Eddington's Principe observation and the quote from Eddington are from Michael Brooks, *Free Radicals: The Secret Anarchy of Science* (London: Profile Books, 2011), pp. 66–68.

5. General relativity's explanation of the discrepancy in Mercury's orbit is described in Bernard Schutz, *Gravity from the Ground Up* (Cambridge: Cambridge University Press, 2003), pp. 49, 237.

6. Einstein's reaction to the discovery that general relativity explained the discrepancy in Mercury's orbit is described in Abraham Pais, *Subtle is the Lord* (Oxford: Oxford University Press, 1983), p. 253.

7. The influence of special and general relativity on the clocks on the GPS satellites is described in Brian Cox and Jeff Forshaw, *Why Does E*=mc²*?* (Cambridge: Da Capo Books, 2009), p. 235.

8. The observation that Einstein did not think much about general relativity between 1907 and 1911, and the use of his letters to underline this is from Abraham Pais, *Subtle is the Lord* (Oxford: Oxford University Press, 1983), pp. 187–90.

9. Einstein's observation shortly after returning to Zurich that he had found "the most general equations" is from Abraham Pais, *Subtle is the Lord* (Oxford: Oxford University Press, 1983), p. 211.

10. Einstein's remark that his cousin drew him to Berlin is quoted in Abraham Pais, *Subtle is the Lord* (Oxford: Oxford University Press, 1983), p. 241.

11. The detail of Einstein's field equations and how they were derived is from Bernard Schutz, *Gravity from the Ground Up* (Cambridge: Cambridge University Press, 2003), pp. 240–60.

12. Frame dragging and its relevance to time travel is described in Brian Clegg, *How to Build a Time Machine* (New York: St. Martin's Press, 2011), pp. 231–34.

13. *The Guinness Book of Records*'s note of the Gravity Probe B spheres is at http://www.guinnessworldrecords.com/Search/Details/Most-spherical-manmade-object/59904.htm.

14. The results of the Gravity Probe B investigation of frame dragging were accepted but not yet published at the time of writing as C. W. F. Everitt, "Gravity Probe B: Final results of a space experiment to test general relativity," *Physical Review Letters.* Accepted May 1, 2011.

15. The laser rangefinder experiment that beat Gravity Probe B to demonstrating frame dragging is described in I. Ciufolini and E. C. Pavlis, "A confirmation of the general relativistic prediction of the Lense-Thirring effect," *Nature*, 431, (October 21, 2004), pp. 958–60.

16. The application of the equivalence principle to a helium balloon in a decelerating car is described in Bernard Schutz, *Gravity from the Ground Up* (Cambridge: Cambridge University Press, 2003), p. 74.

CHAPTER 8

1. Michael Faraday's experimental diary entries on a possible relationship between gravity and electromagnetism are quoted in George Gamow, *Gravity* (New York: Dover, 2002), p. 135.

2. The idea that without neutron stars we would not exist is from Bernard Schutz, *Gravity from the Ground Up* (Cambridge: Cambridge University Press, 2003), p. 266.

3. Information on John Michell is from Brian Clegg, *Before the Big Bang* (New York: St. Martin's Press, 2009), pp. 204–5.

4. The statement that getting away from Earth requires a launch speed of 11.2 kilometers per second is from Bernard Schutz, *Gravity from the Ground Up* (Cambridge: Cambridge University Press, 2003), p. 51.

5. John Michell's paper with the suggestion that big enough stars would not let light out was John Michell, *On the Means of Discovering the Distance, Magnitude, &c. of the Fixed Stars, in consequence of the Diminution of the Velocity of their Light, in case such a Diminution should be found to take place in any of them, and such other Data should be procured from Observations, as would be farther necessary for that Purpose.* Philosophical Transactions of the Royal Society of London, 74 (1783), p. 35.

CHAPTER 9

1. Martin Bojowald's comments on the implications of infinity for a theory are from Martin Bojowald, *Once Before Time: A Whole Story of the Universe* (New York: Alfred A. Knopf, 2010), pp. 4–5.

2. Information on quantum space-time—its implications and—possible means of detection based on variable speed of light are from Shahn Majid, "Quantum Spacetime and Physical Reality" in Shahn Majid (ed.), *On Space and Time* (Cambridge: Cambridge University Press, 2008), pp. 56–140.

3. Details of the standard model of particle physics are from Brian Cox and Jeff Forshaw, *Why Does E*=mc^2*?* (Cambridge: Da Capo Books, 2009), p. 182ff.

4. Martin Bojowald's quip that string theory is a theory of everything because everything and anything can happen is from Martin Bojowald, *Once Before Time: A Whole Story of the Universe* (New York: Alfred A. Knopf, 2010), p. 83.

5. The comment that string and M-theorists could be "away in Never-Never Land" is from Paul Davies, *The Goldilocks Enigma* (London: Penguin, 2007), p. 47.

6. Information on Loop Quantum Gravity is from Martin Bojowald, *Once Before Time: A Whole Story of the Universe* (New York: Alfred A. Knopf, 2010), p. 83ff.

7. Martin Bojowald's movie frame removal analogy for the big bang is in Martin Bojowald, *Once Before Time: A Whole Story of the Universe* (New York: Alfred A. Knopf, 2010), pp. 114–15.

8. The suggestion that loop quantum gravity may require left-handed photons is from Amanda Gefter, "Gravity's bias for left may be writ in the sky," *New Scientist*, March 2, 2011.

9. The suggestion that polarization of the cosmic microwave background could indicate the handedness of gravitons is made in a paper by João Magueijo and Dionigi Benincasa due to be published in 2011 in *Physical Review Letters*.

10. Peter Hořava's challenge to the existence of space-time, producing a new take on quantum gravity is described in Anil Ananthaswamy, "Rethinking Einstein: the end of space-time," *New Scientist*, August 9, 2010.

11. Mark Hadley's thesis was *A Gravitational Theory of Quantum Mechanics*, University of Warwick (UK), December 1996.

12. Details of Mark Hadley's theory are from Marcus Chown,

The Universe Next Door (London: Headline Review, 2003) pp. 63–81.

13. The suggestion of using the distribution of the elements formed in the big bang to provide information on the nature of quantum gravity is from Martin Bojowald, *Once Before Time: A Whole Story of the Universe* (New York: Alfred A. Knopf, 2010), pp. 162–63.

14. Information on the 2005 MAGIC reading that seemed to indicate quantized space is from Anil Ananthaswamy, "Late light reveals what space is made of," *New Scientist,* August 12, 2009.

15. The experiment demonstrating the quantized nature of gravity is from Valery V. Nesvizhevsky, Hans G. Börner, Alexander K. Petukhov, Hartmut Abele, Stefan Baeßler, Frank J. Ruess, Thilo Stöferle, Alexander Westphal, Alexei M. Gagarski, Guennady A. Petrov, and Alexander V. Strelkov, "Quantum states of neutrons in the Earth's gravitational field," *Nature,* 415 (January 17, 2002), pp. 297–99.

16. The more recent quantum gravity experiment enabling very precise measures of the influence of gravity on a quantum particle is described in Tobias Jenke, Peter Geltenbort, Hartmut Lemmel, and Hartmut Abele, "Realization of a gravity-resonance-spectroscopy technique," *Nature Physics,* 7 (2011), pp. 468–72.

17. The suggestion that quantum gravity over very small distances is effectively two dimensional is described in Steven Carlip, *The Small Scale Structure of Spacetime,* arXiv:1009.1136v1, September 6, 2010.

CHAPTER 10

1. Einstein's 1918 development of the theoretical basis for gravitational waves is from George Gamow, *Gravity* (New York: Dover, 2002), p. 141.

2. Information on the early development of gravity wave detectors is from Bernard Schutz, *Gravity from the Ground Up* (Cambridge: Cambridge University Press, 2003), pp. 325–27.

3. Joseph Weber's use of time shifting one of the signals to deduce the rate of random coincidence is described in Harry Collins, *Gravity's Ghost: Scientific Discovery in the Twenty-first Century* (Chicago: University of Chicago Press, 2011), p. 9.

4. The suggestion that adding extra gravity wave detectors to a network could result in doubling and would eventually produce 10 times the detection capability is from Bernard F. Schutz, "Networks of gravitational wave detectors and three figures of merit," *Classical and Quantum Gravity,* vol. 28 (2011), 125023.

5. Harry Collins's strange defense of the failure of gravity wave detection to date is in Harry Collins, *Gravity's Ghost: Scientific Discovery in the Twenty-first Century* (Chicago: University of Chicago Press, 2011), pp. 3–4.

6. The injection of a fake signal to test the response of the detection teams is the main theme of Harry Collins, *Gravity's Ghost: Scientific Discovery in the Twenty-first Century* (Chicago: University of Chicago Press, 2011), all pp.

7. The quote from the scientist about the effect of blind injections is from Harry Collins, *Gravity's Ghost: Scientific Discovery in the

Twenty-first Century (Chicago: University of Chicago Press, 2011), p. 64.

CHAPTER 11

1. The antigravity material cavorite appears in H. G. Wells, *The First Men in the Moon* (London: Collins Library of Classics, undated), all pp.

2. Eric Laithwaite's 1974 Royal Institution lecture on gyroscopes can be seen at http://www.gyroscopes.org.

3. For more on the physics of gyroscopes and Laithwaite's effects see http://www2.eng.cam.ac.uk/~hemh/gyroscopes/html gyroscopes.html.

4. For more on Thomas Townsend Brown's electrogravitic theory linking electromagnetism and gravitation, plus one of the better "Einstein was wrong" attempts to provide an alternative to general relativity see Paul A. LaViolette, *Secrets of Antigravity Propulsion* (Rochester: Bear & Company, 2008), all pp.

5. Basic biographical information on Thomas Townsend Brown is from Nick Cook, *The Hunt for Zero Point* (London: Arrow Books, 2002), p. 31ff.

6. For more on Tesla's ideas on the uses of high-voltage electricity see Brian Clegg, *How to Build a Time Machine* (New York: St. Martin's Press, 2011), pp. 171–75.

7. Evidence that the B2 bomber may use electrostatic technology to enhance stealth/flight (but not for antigravity) is described in

Nick Cook, *The Hunt for Zero Point* (London: Arrow Books, 2002), p. 194.

8. There is some dispute over whether or not Arnold described his UFO as saucer-shaped, but he certainly did make the comment about them moving like a skipping saucer. The term flying saucer seems to originate from newspaper headlines, e.g. "Pilot is baffled by flying 'saucers,'" *Hartford Courant*, June 26, 1947.

9. The article promising the wonders of antigravity including attaining the speed of light is quoted (and pictured) in Nick Cook, *The Hunt for Zero Point* (London: Arrow Books, 2002), p. 3. Cook does not give a source, but refers to a handwritten date of 1956 and the author Michael Gladych.

10. Professor Subrata Roy of the University of Florida's idea for a plasma-supported flying saucer are described in "The World's First Flying Saucer: Made Right Here on Earth," Larry Greenemeier, *Scientific American*, July 7, 2008.

11. One description of German Second World War anti-aircraft directed energy weapons, taken from a U.S. military report is given in Nick Cook, *The Hunt for Zero Point* (London: Arrow Books, 2002), p. 100.

12. The work of the British Radio Research Station is described in Brian Clegg, *Inflight Science* (London: Icon Books, 2011), p. 29.

13. Yevgeny Podkletnov's findings of reduction in gravitational pull over a rotating superconductor is described in Robert Matthews, "Antigravity machine weighed down by controversy," *New Scientist*, September 21, 1996.

14. Reports of Boeing, BAE Systems, and NASA investigating

Podkletnov's effect are covered on the BBC News site in "Boeing tries to defy gravity," July 29 2002, http://news.bbc.co.uk/1/hi/sci/tech/2157975.stm.

15. The inability of current theories to rule out anomalous behavior of antimatter in gravitational fields is described in Michael Martin Neito and T. Goldman "The arguments against 'antigravity' and the gravitational acceleration of antimatter," *Physics Reports,* 205, issue 5 (1991), pp. 221–81.

16. The suggestion that general relativity predicts that antimatter will be repelled by matter (and vice versa) is from M. Villata, "CPT symmetry and antimatter gravity in general relativity," *Europhysics Letters,* 94, 20001 (2011).

17. The idea that antimatter having negative mass would mean changes to the principle of equivalence is discussed in George Gamow, *Gravity* (New York: Dover, 2002), p. 146.

CHAPTER 12

1. Richard Feynman's frank admittance that no one understands quantum physics is from Richard Feynman, *QED: The Strange Theory of Light and Matter* (London: Penguin Books, 1990), p. 9.

2. Richard Feynman's assertion that it doesn't matter if you like a theory is from Richard Feynman, *QED: The Strange Theory of Light and Matter* (London: Penguin Books, 1990), p. 10.

INDEX